Nature, Literature and Culture

Nature, Literature and Culture: Affinities, Influences, Lives and Ideas explores the profound connections between scientific inquiry and literary expression across generations of thinkers and writers.

In poetry, Erasmus Darwin attempts to decode the mysteries of botany and geology, while through autobiography and biography, Charles Darwin makes sense of both himself and his grandfather's shared fascination with biology. Jane Gray concentrates on the life of Asa Gray, a champion of Darwin at Harvard. Both Charles Darwin and Jane Gray humanize their subjects while simultaneously revealing their own perspectives and passions. The Thaxters demonstrate how artistic and interests flow from one generation to the next. Writing humanizes mathematics, physics, chemistry and biology through the works of Marie Curie reflecting on Pierre Curie and herself, Bertrand Russell examining his own life and contributions, and Sara Turing chronicling the remarkable life of her son, Alan Turing. Denis Kelly offers a unique dual perspective as he represents both himself and Winston Churchill, focusing on the intersection of historical life, scholarship and leadership.

This volume concludes with contemporary ecological writing, reflecting our generation's urgent focus on climate and environmental concerns, bringing the conversation full circle from scientific wonder to responsibility.

Jonathan Locke Hart received his PhD in English from the University of Toronto and a PhD in History from the University of Cambridge. He is a fellow of the Royal Society of Canada and member of Academia Europea. He is also senior fellow, Centre for Reformation and Renaissance Studies, Victoria College, University of Toronto; associate at Harvard University Herbaria; life member of Clare Hall, University of Cambridge; senior fellow of Abigail Adams Institute; and Adjunct Professor of Amity School of Languages, Amity University Jaipur. In recent years, he was core faculty of comparative literature at Western University; Chair Professor of the School of Foreign Languages; Director at the Centre for Creative Writing, Literary

Culture and Translation, Shanghai Jiao Tong University; Chair Professor of the School of Translation Studies; and Director of the International Cooperation Centre for Cross-Cultural Studies and Digital Humanities, Shandong University. He has written over thirty books, edited other books and contributed book chapters. A winner of many international awards, including two Fulbrights to Harvard, and having served on national and international committees, including Fulbright and Killam, he has written over a hundred articles and essays and has held visiting appointments at Harvard, Cambridge, Princeton, the Sorbonne-Nouvelle (Paris III), Leiden, UC Irvine, Peking University and elsewhere and has given classes, talks, readings and lectures internationally.

Routledge Interdisciplinary Perspectives on Literature

181 **Narratives of Hope and Despair**
Ruin and Regeneration in Literature and Culture
Edited by Johanna M. Wagner, Melanie Duckworth and Deanna Benjamin

182 **Yuan Zhen's New Music Bureau Poetry**
Music and Ritual as Means of Governance
Mei Ah Tan

183 **Narrating Multiplicities in Motion**
The Transformative Power of Storytelling in Anglophone Cultures
Edited by Elena Ogliari, Anna Pasolini, and Carla Tempestoso

184 **Negative Empathy in Literature and the Arts**
Stefano Ercolino and Massimo Fusillo

185 **Anamorphic Distortion in Literature, Visual Art and Film**
Tony Fabijančić

186 **A New Approach to Reading Videogames as Story**
Virtual Cathedrals in an Age of Decadence
Thomas Rowland

187 **Nature, Literature and Culture**
Affinities, Influences, Lives and Ideas
Jonathan Locke Hart

For more information about this series, please visit: www.routledge.com/Routledge-Interdisciplinary-Perspectives-on-Literature/book-series/RIPL

Nature, Literature and Culture

Affinities, Influences, Lives and Ideas

Jonathan Locke Hart

Routledge
Taylor & Francis Group
NEW YORK AND LONDON

First published 2026
by Routledge
605 Third Avenue, New York, NY 10158

and by Routledge
4 Park Square, Milton Park, Abingdon, Oxon, OX14 4RN

Routledge is an imprint of the Taylor & Francis Group, an informa business

Library of Congress Cataloging-in-Publication Data
Names: Hart, Jonathan Locke, 1956– author
Title: Nature, literature and culture : affinities, influences, lives and ideas / Jonathan Locke Hart.
Description: New York, NY : Routledge, 2026. |
Series: Routledge interdisciplinary perspectives on literature | Includes index. |
Identifiers: LCCN 2026006639 (print) | LCCN 2026006640 (ebook) |
ISBN 9781041098102 hardback | ISBN 9781041098119 paperback |
ISBN 9781003651871 ebook
Subjects: LCSH: Nature in literature | Nature writers | Ecoliterature—History and criticism |
LCGFT: Literary criticism
Classification: LCC PN48 .H37 2026 (print) | LCC PN48 (ebook) | DDC 809/.9336—dc23/eng/20260325
LC record available at https://lccn.loc.gov/2026006639
LC ebook record available at https://lccn.loc.gov/2026006640

ISBN: 978-1-041-09810-2 (hbk)
ISBN: 978-1-041-09811-9 (pbk)
ISBN: 978-1-003-65187-1 (ebk)

DOI: 10.4324/9781003651871

Typeset in Sabon
by codeMantra

For Donald H. Pfister and the Harvard University Herbaria

Contents

Preface and Acknowledgements

Nature, Literature and Culture: Affinities, Influences, Lives and Ideas is about topics long close to my heart. Evidence and close interpretation are important matters in science, the social sciences and the humanities. The book involves close reading of texts by or about those interested in botany, biology, chemistry, mathematics, logic, physics, literature and culture.

I wish to thank those I have thanked in previous books as well as the Harvard University Herbaria; the Faculty Deans, staff and students, Kirkland House, Harvard; the Departments of English and Comparative Literature at Harvard; Director, fellows and students at the Abigail Adams Institute; President, fellows, students and staff of Clare Hall, Cambridge; the master, fellows, students and staff of Churchill College, Cambridge; Director and staff, Churchill Archives Centre; Director and staff of the Centre Culturel Irlandais, Paris; and my colleagues and students in Comparative Literature at Sorbonne Nouvelle (Paris III) and at Princeton University, Shanghai Jiao Tong University, Peking University, Shandong University and elsewhere. Regarding *Nature, Literature and Culture*, I thank Donald H. Pfister and the Harvard University Herbaria for the chapters on the Grays and the Thaxters, the work on the Grays leading me back to Cambridge and England and to Darwin and his family. I remember with gratitude Priscilla L. Walton, who helped make possible the original publication of my work on the Grays.

My work on the Curies comes out of my time teaching a doctoral seminar on otherness in Comparative Literature in the old Sorbonne building, and my thanks to Jean Bessière, my colleagues and my students at the Sorbonne Nouvelle, and to Nora Hickey M'Sichili, Yann Le Cadre, Marion Mossu, Patricia Costello and other colleagues, students and friends at the Centre Culturelle Irlandais. With thanks, I remember my father for encouraging my early interest in Bertrand Russell and my Logic Professor John G. Slater, a scholar of Russell. My work on Alan Turing comes, in part, from discussions and working with my friend and colleague, Anders Lindquist, Chair Professor in Mathematics and Engineering, at Shanghai

Jiao Tong University, where we both taught. Allen Packwood, Director of the Churchill College Archives, kindly encouraged and enabled my study of Winston Churchill and Denis Kelly, and I also thank Victoria Kelly for permission to discuss her father. I express my gratitude for the Jennie Churchill Award and thank the master, fellows and scholars of Churchill College and the team at the Churchill Archives Centre. The next chapter goes back to my early engagement with ecology, formally in the 1970s when I worked in the wilderness, including as a junior forest ranger, but my thanks for encouraging and publishing this work to Priscilla L. Walton (in memory) and Christopher Vanderwees. Scholarship is a cooperative activity and I owe a debt to colleagues, students, librarians, archivists and others. More generally, I thank David Porter and Wu Shang for their kindness, advice and help. My gratitude to Anthony Pagden, an exemplary teacher, scholar and writer, whom I was fortunate enough to have as my supervisor at Cambridge.

Thanks to the Director of the Centre for Reformation and Renaissance Studies (CRRS), Victoria College, the University of Toronto, Misha Teramura, to the former Director, Ethan Matt Kaveler, the Assistant to the Director, Dr Natalie Oeltjen, the fellows and Konrad Eisenbichler and James Estes.

Thanks to the Director, Danilo Petranovich, and the fellows and students of Abigail Adams Institute; the faculty and students of Amity University Jaipur, Rajasthan, including Dipa Chakrabarti and Gunja Patni; and to the University of Macau, especially to Li Defeng, and to the master, fellows, staff and students at Choi Kai Yau College.

My thanks again to Routledge, to my editor, Jennifer Abbott, for her kindness and support for many books till now and to their colleagues, including Anita Bhatt and Vaishnavi Singh, editorial assistants. I also owe a debt to the reviewers of the manuscript and to the scholars who helped me. Moreover, my gratitude to the anonymous readers for their constructive comments and perceptive readings.

I thank librarians and archivists with whom I have worked regularly on this and other work in many libraries and archives in France, Britain, the United States, Canada and China.

Thanks to my friends and colleagues, including Nick Barraud, Tom and Verena Conley, Merrill Distad, Johanne Du Berger, William Fleury, Luis Galvan, Colin Hill, Timothy Kilbourn, Nicholas Korsakov, Kevin Lewis, Li Defeng, Hao Li, Anders Lindquist, Linda Matazzaro, Stephen Mobbs, Cathleen Pfister, Jack Robinson, Glenn Rollans, Eric Schloss, Jüri Talvet, Andrew Taylor, Pauline Thomas, Nadezda Vashkevich, Ming Xie and I-Chun Wang.

Finally, to my wife, Mary, and our children, Julia and James; to Lolita Cayanan as well as to Lynn Harris; and my brothers and sisters, Charles, Gwendolyn, Deborah and Jennifer. I remember my father, George Edward, my mother, Jean Jackman and my brother Alan. In addition to my nieces and nephews, I thank my cousins, including Betty Bednarski, Nora Down and Paul Durber.

Introduction

It may be something that is forgotten in our culture, that is, scholars and scientists writing for a wider audience. I begin with a couple of passages from Bertrand Russell, something that humanists might remember and perhaps part of my apology in advance for, once more, trying to write for general readers, non-specialists and students and not simply for specialists in my own areas of research. First, Russell says:

> I suggest to young professors that their first work should be written in a jargon only to be understood by the erudite few. With that behind them, they can ever after say what they have to say in a language 'understanded of the people'.[1]

Russell, a mathematician, logician and philosopher, writes in a way that is accessible. I strive to do the same. Second, Russell observes: "I am not thinking of what history does for historians; I am thinking of history as an essential part of the furniture of an educated mind."[2] I think of history, literature, culture, humanities, social science and science as part of that furniture. Educating our minds, our reason and imagination should be for as many people seek that education.

Nature, Literature and Culture has several threads running through it – the relation between poetry and literature on the one hand and nature (including botany and natural selection) on the other, the role of representation and self-representation of lives and the role of the writer and family (for instance Erasmus Darwin, Charles Darwin, the Grays, the Thaxters, Bertrand Russell, Denis Kelly and Winston Churchill), affinities and influences (the Darwins, the Grays), texts and contexts more generally. The chapters often involve the literary representation of affinities, influences, lives and ideas. For instance, in poetry, Erasmus Darwin tries to make sense of botany and geology, and through autobiography and biography, Charles Darwin makes sense of himself and his grandfather and their interest in biology; Jane Gray concentrates on the life of Asa

DOI: 10.4324/9781003651871-1

Gray, a champion of Darwin at Harvard and in Boston and the United States. Charles Darwin and Jane Gray humanize their subjects and reveal themselves. The Thaxters show an interest in the arts and nature from generation to generation. Writing humanizes mathematics, physics, chemistry and biology in the work of Marie Curie on Pierre Curie and herself, Bertrand Russell on his life and work, Sara Turing and others on her son, Alan Turing. After the Second World War, Denis Kelly, represents himself as he represents Winston Churchill, whose assistant he was, with a focus on a life of history, of a historian and leader. The final chapter is a discussion of three works on ecology written in recent years, climate being a focus in this generation. The social and cultural concerns can also involve writing about conflict, empire and its aftermath, the relations between European (settler) and Indigenous peoples. The world has literary, historical, theoretical and other writing to make sense of nature and culture and that writing ties the book together.

From the birth of Darwin to the present, words matter. Even mathematics, logic and physics can be discussed in words as the chapters on the Curies, Russell and Turing illustrate. Central to this book is mimesis and auto-mimesis, representation and self-representation, text and context, story and history, science and culture, narrative and argument (also what I have called story-argument elsewhere). Rather than repeat my earlier work, I mention it.[3] One of the central points of *Nature, Literature and Culture* is how scientists represent themselves and their families, and how families represent them is a key for understanding science and scientists and their contexts. This point applies to the Darwins, Grays, Thaxters and Turings. Bertrand Russell writes about himself and the world, about mathematics and logic and about social and historical matters. Darwin appears in the chapter on the Grays and Asa Gray appears in the chapter on the Thaxters and Bertrand Russell is alluded to in the section on the Turings. These representations and self-representations humanize science and scientists, both of which are so important to our lives and society, to our ideas, culture, economy and so much more. I studied history of ideas (intellectual history), including the history of science, and others have called me an intellectual and cultural historian. I respect evidence and testing ideas and theories, and see that as a goal of humanities and social sciences as well as the method of science; however, paradigms shift, to echo Thomas Kuhn.[4] My own emphasis is on self-expression and freedom of expression of scientists, humanists, poets, family members and others. Science has a context. The personal and the individual matter.

The book is mainly chronological. It begins about the time of Darwin's birth, examines his writings about his life and his grandfather Erasmus Darwin then proceeds to discuss Jane Gray's discussion of her husband,

Asa Gray, who helped to bring Darwin's ideas to Harvard and the United States, then examines Levi, Celia and Roland Thaxter, also a Harvard family with interests in science and the arts, then analyses Marie Curie's biography of Pierre Curie and her autobiographical notes and surrounding texts, then Bertrand Russell's writings on writing, his life, history, society, mathematics, logic science and other topics, then focuses on the representation of Alan Turing's life and his work in mathematics and related areas by his mother Sara Turing while also discussing the representation of Alan by others, principally his brother, John Turing, but also by Andrew Hodges and Martin Davis, both having worked in areas of mathematics and physics that allow them to understand Alan's contribution, then, shifts to the world after the Second World War, explores the relation between Denis Kelly and Winston Churchill, assistant and the historian, writer and statesman and finally, in the body of the book, concentrates on Alexander Menrisky, Juliana Chow on ecology natural history and Lowell Wyse on ecospatiality in American literature respectively.[5] There are themes that run through my book, for instance, imperialism and freedom of expression in Darwin, Russell, Kelly, Churchill and others. The role of women is significant, as can be seen, for instance, in the discussions of Jane Gray, Celia Thaxter, Marie Curie and Sara Turing. Jane, Marie and Sarah representing their families, and Celia is a poet. Representation is also self-representation to some extent. Other women are important in the book. The dignity of the individual and the importance of nature and the natural world, humans and the human world underlie my study. The elements can be observed in Darwin's love of observing the natural world and his opposition to slavery; Jane Gray's representation of domestic life; Celia Thaxter's love of her garden and her poetry about nature; Marie Curie's view of Pierre and their children as well as science; Bertrand Russell's commitment to human knowledge and freedom; Alan's Turing's love of mathematics, computers and running and his mother standing by him in death; Denis Kelly's sympathetic representation of Winston Churchill as a man and historian; the nature, ecology and related subjects in an industrialized world. From natural history to natural history then and now, the book begins and ends.

The method I use is one of close attention to textual evidence and close reading. I examine books and texts about nature, literature and culture and try to pay attention to all texts. C. P. Snow, trained in science and a novelist, spoke of the two cultures.[6] I hope that one contribution of this book is to cross boundaries and speak of science, literature and culture as related aspects of our lives and society. Words, texts, works matter. From Darwin through the Grays and Thaxters, the Curies and Bertrand Russell to Alan Turing and beyond, science and the arts and humanities are intertwined, part of a life of culture and an educated mind. It is contingent on those of us who are not scientists to understand science and scientists, to respect nature and science and not literature and culture alone.

This book comes out of my interest in science and the history of science, to my good luck to have taught comparative literature, English and history and to be affiliated with a faculty of medicine, interdisciplinary centres and institutes and the Harvard University Herbaria. My life as a student, scholar and visitor, in colleges, for instance at Toronto Oxford, Cambridge, Harvard and Princeton, involved a community of scholars and students across the fields. *Nature, Literature and Culture* is one of my attempts to contribute to science, which is too often precarious because reason and a study of nature sometimes seem to be a threat, not to all but to many. It is harder to build up respect, rights and dignity for the individual than to tear it down. As a society, toleration, openness, innovation, knowledge, kindness and respect are needed for individuals and society. Cooperation among individuals is necessary for peace, harmony and well-being. None of this is easy, and one can only hope and work. Free speech and free inquiry, as vulnerable as they are, are precious. There is no preaching about any of this. It is really a matter for the voices of the people I discuss to reveal themselves, others and the world. I turn to them now.

Notes

1 Bertrand Russell, "How I Write," in *The Basic Writings of Bertrand Russell*, edited by Bertrand Russell (1st ed.) (London: Routledge, 2009), 37.

2 Bertrand Russell, "History as an Art," in *The Basic Writings*, edited by Bertrand Russell, 515.

3 For story-argument, Jonathan Locke Hart, "Stephen Greenblatt's Shakespearean Negotiations," *Textual Practice* 5 (1991): 429–448. On mimesis, see, for instance, Jonathan Hart, *Representing the New World: The English and French Uses of the Example of Spain, 1492–1713* (New York: Palgrave, 2001); Jonathan Hart, *Textual Imitation: Making and Seeing in Literature* (New York: Palgrave, 2013); Jonathan Locke Hart, *Aristotle and His Afterlife: Rhetoric, Poetics and Comparison* (Paris: Honoré Champion, 2019) and Jonathan Locke Hart, *Making and Seeing Modern Texts* (New York and London: Routledge, 2019). On science and culture, Jonathan Hart, see "Introduction," in *City of the End of Things: Lectures on Civilization and Empire*, edited by Jonathan Hart (Toronto: Oxford University Press, 2009), 1–34. On ecology and eco-imperialism, see Jonathan Hart, *Empires and Colonies* (Cambridge: Polity Press, 2008), a volume in their series, "Themes in History."

4 Thomas S. Kuhn, *The Structure of Scientific Revolutions* (Chicago, IL: University of Chicago Press, 1962); Thomas S. Kuhn, *The Essential Tension : Selected Studies in Scientific Tradition and Change* (Chicago, IL: University of Chicago Press, 1977); Thomas S. Kuhn, *The Road since Structure : Philosophical Essays, 1970–1993, with an Autobiographical Interview*, edited by James Conant and John Haugeland (Chicago, IL: University of Chicago Press, 2000). See Paul Hoyningen-Huene, *Reconstructing Scientific Revolutions: Thomas S. Kuhn's*

Philosophy of Science (Chicago, IL: University of Chicago Press, 1993); Pawel Jarnicki and Hajo Greif, "The 'Aristotle Experience' Revisited: Thomas Kuhn Meets Ludwik Fleck on the Road to Structure," *Archiv Für Geschichte Der Philosophie* 106 (2) (2024): 313–349.

5 Juliana Chow, *Nineteenth-Century American Literature and the Discourse of Natural History* (Cambridge: Cambridge University Press, 2022); Alexander Menrisky, *Wild Abandon: American Literature and the Identity Politics of Ecology* (Cambridge: Cambridge University Press, 2021); Lowell Wyse, *Ecospatiality : A Place-Based Approach to American Literature* (Iowa City: University of Iowa Press, 2021).

6 C. P. Snow, *The Two Cultures and the Scientific Revolution* (Cambridge: Cambridge University Press, 1959); C. P. Snow, *The Two Cultures; and, A Second Look: An Expanded Version of "The Two Cultures and the Scientific Revolution."* (London: Cambridge University Press, 1969); See also F. R. Leavis and Michael Yudkin, *Two Cultures? : The Significance of C. P. Snow. Being the Richmond Lecture, 1962. With a New Pref. for the American Reader. And an Essay on Sir Charles Snow's Rede Lecture* (New York: Pantheon Books, 1963); Anthony J. Wiener, "Introduction to the Supplement to This Special Issue on C. P. Snow's 'Two Cultures': A Case of Morality, Organ Transplants, and a Third Culture," *Technology in Society* 32 (1) (2010): 53–55; Struan Jacobs, "C. P. Snow's the Two Cultures: Michael Polanyi's Response and Context," *Bulletin of Science, Technology & Society* 31 (3) (2011): 172–78; Rupert Cole, "A Tale of Two Train Journeys: Lawrence Bragg, C. P. Snow and the 'Two Cultures,'" *Interdisciplinary Science Reviews* 41 (2–3) (2016): 133–147; W. Patrick McCray, "Snow's Storm the Two Cultures and the Scientific Revolution C. P. Snow Cambridge Univ. Press, 1959. 52 Pp," *Science (American Association for the Advancement of Science)* 364 (6439) (2019): 430–432.

1 Charles Darwin
Youth

Charles Darwin wrote an autobiography and a biography of his grandfather, Erasmus Darwin. The biography was published while Darwin was alive, but it grew out of his autobiography which was published posthumously. The biography of some of the extended Darwin family seemed to hope would rehabilitate the reputation of Erasmus Darwin. The family of Charles Darwin edited his work posthumously, his son Francis, soon after the death of Darwin and his granddaughter Nora Barlow in 1958.[1] I take the words of these editors who were family members, but my focus is really on what Darwin says about science and culture in relation to himself and to his grandfather Erasmus.

Francis included a note when the autobiography was published:

> MY father's autobiographical recollections, given in the present chapter, were written for his children,—and written without any thought that they would ever be published. To many this may seem an impossibility; but those who knew my father will understand how it was not only possible, but natural. The autobiography bears the heading, 'Recollections of the Development of my Mind and Character,' and end with the following note:—'Aug. 3, 1876. This sketch of my life was begun about May 28th at Hopedene,* and since then I have written for nearly an hour on most afternoons.' It will easily be understood that, in a narrative of a personal and intimate kind written for his wife and children, passages should occur which must here be omitted; and I have not thought it necessary to indicate where such omissions are made. It has been found necessary to make a few corrections of obvious verbal slips, but the number of such alterations has been kept down to the minimum.
>
> —F. D.[2]

Francis justifies the omissions and shows where they occur. Charles Darwin did not control the editing and publication of his life. The large and talented family of the Darwin-Wedgwoods (and related families) is part of the

DOI: 10.4324/9781003651871-2

theme and the reception of this autobiography and is connected with the writing of Charles Darwin's biography of Erasmus Darwin. In her Preface, Nora Barlow notes this editorial history:

> IN HIS old age Charles Darwin wrote down his recollections for his own amusement and the interest of his children and their descendants. He finished the main narrative of 121 pages between May and August, 1876, writing as he tells us for an hour on most afternoons. During the last six years of his life he enlarged on what he had already written as fresh memories occurred to him, inserting the sixty-seven further pages of Addenda into their appropriate places. The present edition of the *Autobiography* is a complete transcript of the whole manuscript, now housed in its old leather binding in the Cambridge University Library.[3]

By transcribing the manuscript of her grandfather's autobiography, Barlow seeks, in 1958, to restore what Francis Darwin omitted in 1887. Barlow is careful to show how her editing differs from that of Francis Darwin and how she has restored Charles Darwin's text and what she has done to it as an editor:

> I have followed the original closely, restoring omissions amounting to nearly six thousand words, and correcting many trivial errors and alterations that had crept into the earlier rendering; and where necessary I have changed erratic punctuation and filled in purely formal abbreviations, both of which checked smooth reading. Throughout, Charles Darwin's parentheses are in round brackets; my own additions are indicated by square ones. Footnotes in Francis Darwin's edition of the *Autobiography* are initialled F. D., those added by me are initialled N. B. To maintain the continuity of the text, I have not marked the earlier omissions as they occur, but there is a page and line reference to the more important at the end of the book for those who wish to trace them.[4]

Barlow is interested in "smooth reading" and although she gives no examples of what constitute "many trivial errors and alterations," the nature of the mistakes and changes, the readers know that they are not getting Charles Darwin's manuscript as he wrote it and to do that one must go and read the manuscript in the University Library at Cambridge. Barlow (1885–1989) herself has produced something that is closer to that experience than did Francis Darwin (1848–1925), who was guarding the privacy of the family in a publication five years after the death of his father, Charles (1809–1882). Barlow thanks her family, including Sir Charles Darwin and her cousins, and that includes one of the Cornfords, and Dr. Padel. The Darwins and related families had many ongoing connections in many places and at Cambridge, something I experienced when, I seem to remember, Ruth Padel, the granddaughter of Nora Barlow and the great great granddaughter of Charles Darwin, and I read poems at the eightieth birthday celebration of George Steiner at Churchill College, Cambridge. At Cambridge, one was also likely to run into a Huxley. Both Francis Darwin and Nora Barlow seem to have wished to present Charles Darwin to the best avail. I chose Barlow's edition as it is based on a transcript and seems closest to Darwin's manuscript. I cannot expect the reader to be checking from the manuscript, even if and when there is a digitized version. An interesting quotation occurs on page 8 that shows something of Darwin's attitude toward his own work and to science:

> 'If I lived twenty more years and was able to work, how I should have to modify the *Origin*, and how much the views on all points will have to be modified! Well it is a beginning, and that is something ....' *Charles Darwin to J. D. Hooker*, 1869.

Science is about modification, and Darwin understood self-modification and the need in time to work on all points, something that time does not afford the individual scientist the opportunity, and it is left to others to continue the work.

Moreover, my focus is on what Darwin says about culture and science in context of writing about his own life. He begins with a framing of what he intends with his autobiographical account:

> **May 31st, 1876**
>
> **Recollections of the Development of My Mind and Character**
>
> A GERMAN EDITOR having written to me to ask for an account of the development of my mind and character with some sketch of my autobiography, I have thought that the attempt would amuse me, and might possibly interest my children or their children. I know that it would have interested me greatly to have read even so short and dull a sketch of the mind of my grandfather written by himself, and what he thought and did and how he worked. I have attempted to write the following account of myself, as if I were a dead man in another world looking back at my own life. Nor have I found this difficult, for life is nearly over with me. I have taken no pains about my style of writing.[5]

This autobiographical work arises from an invitation, and Darwin accepted it, imagining it as being for his family and saying that he would have been interested in such an account of Erasmus Darwin, his thinking, and his doings, and the way he went about his work. It is interesting that Charles views his life writing as a link between the life of his grandfather and that of his own. Charles imagines as if he were a writer from beyond the grave and sees himself as in the twilight of life. Given the circumstances, Charles Darwin says that he did not attend to style. Time and timing are important. Past, present and future are all part of the thinking and life writing of Charles Darwin.

Charles Darwin is modest about his mind. He tells a number of stories against himself. After Darwin tells of his birth in a sentence, he says:

> I have heard my Father say that he believed that persons with powerful minds generally had memories extending far back to a very early period of life. This is not my case for my earliest recollection goes back only to when I was a few months over four years old, when we went to near Abergele for sea-bathing, and I recollect some events and places there with some little distinctness.[6]

It is as though Darwin is trying to show that, by his father's standards, he did not have a powerful mind. His memory was late and was not entirely distinct. Darwin's work is not based on conceit but perhaps on the conceit of sincerity and modesty or simply his being sincere and modest by nature. He then tries to analyse why he does not remember much about his mother who died when he was eight, and examines this lack of memory to the grief of his sisters. Darwin balances this forgetfulness with his innate qualities:

> By the time I went to this day-school my taste for natural history, and more especially for collecting, was well developed. I tried to make out the names of plants, and collected all sorts of things, shells, seals, franks, coins, and minerals. The passion for collecting, which leads a man to be a systematic naturalist, a virtuoso or a miser, was very strong in me, and was clearly innate, as none of my sisters or brother ever had this taste.[7]

He speaks about taste and innate passion. Darwin balances his strengths and weaknesses. Just when the reader might think that Darwin is going to extol or develop a sense of his innate genius for natural history, he tells another tale against himself:

> One little event during this year has fixed itself very firmly in my mind, and I hope that it has done so from my conscience having been afterwards sorely troubled by it; it is curious as showing that apparently I was interested at this early age in the variability of plants! I told another little boy (I believe it was Leighton, who afterwards become a

> well-known Lichenologist and botanist) that I could produce variously coloured Polyanthuses and Primroses by watering them with certain coloured fluids, which was of course a monstrous fable, and had never been tried by me. I may here also confess that as a little boy I was much given to inventing deliberate falsehoods, and this was always done for the sake of causing excitement. For instance, I once gathered much valuable fruit from my Father's trees and hid them in the shrubbery, and then ran in breathless haste to spread the news that I had discovered a hoard of stolen fruit.[8]

Darwin portrays himself as a boy with a passion for plants but also as a liar and fabulist. His example includes his schoolmate William Allport Leighton (1805–1899), a biologist, also later trained at Cambridge. Barlow includes a note by Francis Darwin and another by herself, so the edition is cumulative and not a contest over editing or representing Charles Darwin.

Darwin continues to tell stories about his childhood, and, except for some details, he makes himself seem like a naïve and unexceptional child. He liked to collect eggs and usually took only one egg from a bird nest, except for once out of bravado, when he took all the eggs. He liked to angle. It is as though sometimes he sees himself as an unattractive character as a child:

> Once as a very little boy, whilst at the day-school, or before that time, I acted cruelly, for I beat a puppy I believe, simply from enjoying the sense of power; but the beating could not have been severe, for the puppy did not howl, of which I feel sure as the spot was near to the house.[9]

Darwin's self-portraits have twists and turns, an almost oxymoronic virtue-vice, fill of ambivalence and contradiction. He continues his narrative:

> This act lay heavily on my conscience, as is shown by my remembering the exact spot where the crime was committed. It probably lay all the heavier from my love of dogs being then, and for a long time afterwards, a passion. Dogs seemed to know this, for I was an adept in robbing their love from their masters.[10]

Darwin's tale of cruelty with his abuse of the puppy turns to his love of dogs and his adeptness at winning their love. It is almost a comic reversal from sadness to joy, from fall to redemption. He went to Dr. Butler's school in Shrewsbury from 1818 to 1825, and its emphasis was on classics. Darwin claims that the experience at school as "a blank" and adds: "During my whole life I have been singularly incapable of mastering any language. Especial attention was paid to verse-making, and this I could never do well."[11] His further description seems to contradict his statement about his lack of facility with languages:

> Much attention was paid to learning by heart the lessons of the previous day; this I could effect with great facility learning forty or fifty lines of Virgil or Homer, whilst I was in morning chapel; but this exercise was utterly useless, for every verse was forgotten in forty-eight hours. I was not idle, and with the exception of versification, generally worked conscientiously at my classics, not using cribs. The sole pleasure I ever received from such studies, was from some of the odes of Horace, which I admired greatly.[12]

It seems that Darwin had a good memory and, presumably, he was reading Homer, Virgil and Horace in the original. Memorization is not something that Darwin seems to have prized. He worked hard except for writing verses. Not much resembled scientific training in this school.

Darwin, who does not seem to wish to represent himself as anything but, in general at least, an ordinary child, turns to a description of his father:

> When I left the school I was for my age neither high nor low in it; and I believe that I was considered by all my masters and by my Father as a very ordinary boy, rather below the common standard in intellect. To my deep mortification my father once said to me, "You care for nothing but shooting, dogs, and rat-catching, and you will be a disgrace to yourself and all your family." But my father, who was the kindest man I ever knew, and whose memory I love with all my heart, must have been angry and somewhat unjust when he used such words.[13]

In this report, the schoolmasters and his father thought Darwin very ordinary intellectually. Family is important to Darwin, and disgracing his family seems to have stung. Darwin still praises his father as "a remarkable man."[14] Darwin says that his father was tall and heavy and his "chief mental characteristics were his powers of observation and his sympathy, neither of which have I ever seen exceeded or even equalled."[15] Darwin sees his father's success as a physician, which he worked at for over sixty years, as coming from that sympathy, and he describes what his father thought of his profession and the role his father's father had in the choice of work:

> His great success as a doctor was the more remarkable, as he told me that he at first hated his profession so much that if he had been sure of the smallest pittance, or if his father had given him any choice, nothing should have induced him to follow it.[16]

Erasmus Darwin, it seems, did not give Darwin's father a choice. Neither Darwin nor his father, Robert Waring Darwin (1766–1848), according to the son, liked the sight of blood. Charles Darwin assesses the intellect and character of his father: "But the most remarkable power which my father possessed was that of reading the characters, and even the thoughts of those whom he saw even for a short time."[17] Darwin, considering his father to have remarkable powers, also gives details that show the intricacy of matters. For instance, Darwin says: "I was told that a young Doctor in Shrewsbury, who disliked my father, used to say that he was wholly unscientific, but owned that his power of predicting the end of an illness was unparalleled."[18] Here, Dr. Robert Darwin has intuition or a special power but is not scientific. Darwin's portrait of his father is detailed. One aspect is memory:

> My father possessed an extraordinary memory, especially for dates, so that he knew, when he was very old the day of the birth, marriage, and death of a multitude of persons in Shropshire; and he once told me that this power annoyed him; for if he once heard a date he could not forget it; and thus the deaths of many friends were often recalled to his mind. Owing to his strong memory he knew an extraordinary number of curious stories, which he liked to tell, as he was a great talker. He was generally in high spirits, and laughed and joked with every one—often with his servants—with the utmost freedom; yet he had the art of making every one obey him to the letter.[19]

This portrait of Darwin's father represents his memory as a blessing and a burden. The father's memory meant that he had many stories that he recounted in conversation. He had a good and open disposition while also commanding obedience. Darwin tells about his father's use of having a patient fear him for compassionate reasons that enhance health and liberty. Despite Darwin's admiration for his father, he is sometimes frank about his father's influence on him:

> My father's mind was not scientific, and he did not try to generalise his knowledge under general laws; yet he formed a theory for almost everything which occurred. I do not think that I gained much from him intellectually; but his example ought to have been of much moral service to all his children. One of his golden rules (a hard one to follow) was, 'Never become the friend of any one whom you cannot respect.'
>
> With respect to my Father's father, the author of the *Botanic Garden* etc., I have put together all the facts which I could collect in his published *Life*.[20]

Here, in his own words, Darwin returns to a view that his father was not scientific, which, for Darwin, is to generalize knowledge into general laws. He says he did not learn much intellectually from his father, but he does think of his father's moral influence. Darwin then thinks of his father's father Erasmus and refers to his own life. Family is a key to Darwin, and he then focuses on his brother and his sisters. By examining his family, Darwin looks for what he has in common with those related to him and what is distinctive about him. Darwin is defining himself through comparison.

Darwin's brief portrait of his brother and his sisters combines affection with frankness:

> My brother Erasmus possessed a remarkably clear mind, with extensive and diversified tastes and knowledge in literature, art, and even in science. For a short time he collected and dried plants, and during a somewhat longer time experimented in chemistry. He was extremely agreeable, and his wit often reminded me of that in the letters and works of Charles Lamb. He was very kind-hearted; but his health from his boyhood had been weak, and as a consequence he failed in energy. His spirits were not high, sometimes low, more especially during early and middle manhood. He read much, even whilst a boy, and at school encouraged me to read, lending me books. Our minds and tastes were, however, so different that I do not think that I owe much to him intellectually—nor to my four sisters, who possessed very different characters, and some of them had strongly marked characters. All were extremely kind and affectionate towards me during their whole lives.[21]

Darwin discusses his brother's sense of literature and science, including reading and collecting, but whatever positive feelings he represents that he has for his brother and sisters, who are unnamed, he also addresses their influence on him. As with Darwin's father, with his siblings, Darwin thinks he does not owe much to them intellectually. Darwin accepts some of his cousin's ideas: "I am inclined to agree with Francis Galton in believing that education and environment produce only a small effect on the mind of anyone, and that most of our qualities are innate."[22] Galton himself seems to have balanced environment with innate qualities more than Darwin does here. Their relation was intricate.[23] Family is important to Darwin's view of his own development and his own view of character in others and himself:

> Looking back as well as I can at my character during my school life, the only qualities which at this period promised well for the future,

> were, that I had strong and diversified tastes, much zeal for whatever interested me, and a keen pleasure in understanding any complex subject or thing. I was taught Euclid by a private tutor, and I distinctly remember the intense satisfaction which the clear geometrical proofs gave me. I remember with equal distinctness the delight which my uncle gave me (the father of Francis Galton) by explaining the principle of the vernier of a barometer. With respect to diversified tastes, independently of science, I was fond of reading various books, and I used to sit for hours reading the historical plays of Shakespeare, generally in an old window in the thick walls of the school. I read also other poetry, such as the recently published poems of Byron, Scott, and Thomson's *Seasons*.[24]

Darwin speaks of his own "zeal" and "keen pleasure" in comprehending intricate things and subjects. Euclid brought Darwin "satisfaction" and the barometer brought him "delight." Besides family, school and reading are topics that Darwin discusses. Poetry was something he enjoyed. Darwin also records the changes in his own tastes:

> I mention this because later in life I wholly lost, to my great regret, all pleasure from poetry of any kind, including Shakespeare. In connection with pleasure from poetry I may add that in 1822 a vivid delight in scenery was first awakened in my mind, during a riding tour on the borders of Wales, and which has lasted longer than any other aesthetic pleasure.[25]

Darwin shifts to a sense of aesthetics from poetry to landscape when he mentioned a trip he took in his youth. Culture yields to nature. Still, Darwin discusses reading:

> Early in my school-days a boy had a copy of the *Wonders of the World*, which I often read and disputed with other boys about the veracity of some of the statements; and I believe this book first gave me a wish to travel in remote countries, which was ultimately fulfilled by the voyage of the *Beagle*. In the latter part of my school life I became passionately fond of shooting, and I do not believe that anyone could have shown more zeal for the most holy cause than I did for shooting birds.[26]

Evelyn M. Witkin has tried to identify the book Darwin is talking about and thinks it is Nathaniel Wanley's *The Wonders of the Little World* (1678).[27] Darwin says that this book affected his wish to travel, which led to his being on the voyage of the *Beagle*. He also raises his "zeal" for shooting birds. In this portrait, the future collector and naturalist also enjoyed killing. These are complex and apparently contradictory impulses. Darwin shifts his emphasis:

> With respect to science, I continued collecting minerals with much zeal, but quite unscientifically—all that I cared for was a new named mineral, and I hardly attempted to classify them. I must have observed insects with some little care, for when ten years old (1819) I went for three weeks to Plas Edwards on the sea-coast in Wales, I was very much interested and surprised at seeing a large black and scarlet Hemipterous insect, many moths (Zygæna) and a Cicindela, which are not found in Shropshire. I almost made up my mind to begin collecting all the insects which I could find dead, for on consulting my sister, I concluded that it was not right to kill insects for the sake of making a collection. From reading White's *Selborne* I took much pleasure in watching the habits of birds, and even made notes on the subject. In my simplicity I remember wondering why every gentleman did not become an ornithologist.[28]

Darwin stresses his lack of science in his collection of minerals as well as his interest in insects, including his sense that killing insects for collecting is wrong. He also watched birds and thought all gentlemen should become ornithologists. From the shooting of birds, Darwin has moved to watching them.

Science is front of mind for Darwin when he remembers his life as a student:

> Towards the close of my school life, my brother worked hard at chemistry and made a fair laboratory with proper apparatus in the tool-house in the garden, and I was allowed to aid him as a servant in most of his experiments. He made all the gases and many compounds, and I read with care several books on chemistry, such as Henry and Parkes' *Chemical Catechism*. The subject interested me greatly, and we often used to go on working till rather late at night. This was the best part of my education at school, for it showed me practically the meaning of experimental science.[29]

This practical experience, which his brother made possible, taught him experimental science. Once more, memories of family Darwin intertwines with thoughts on science. He continues to express his childhood enthusiasm for the various branches of that endeavour. Decisions and fortune in the family affect his studies:

> As I was doing no good at school, my father wisely took me away at a rather earlier age than usual, and sent me (October 1825) to Edinburgh

> University with my brother, where I stayed for two years or sessions. My brother was completing his medical studies, though I do not believe he ever really intended to practise, and I was sent there to commence them. But soon after this period I became convinced from various small circumstances that my father would leave me property enough to subsist on with some comfort, though I never imagined that I should be so rich a man as I am; but my belief was sufficient to check any strenuous effort to learn medicine.[30]

Francis Darwin's note says that records show Darwin and his brother, Erasmus, used the library more than was usual for students of the time.[31] Darwin's father was a well-off doctor, and so his sons did not have to practise medicine even if they studied it. Darwin then mentions that he prefers reading to lectures and regrets he did not get experience in dissection despite his disgust with anatomy. His father, with whom Darwin worked the summer before going to Edinburgh and whom Darwin reiterates is the best judge of character he ever knew, thought Darwin would make a successful doctor. According to Darwin, to excite confidence is the main aspect of success, and Darwin does not know what his father saw in him that made him confident in Darwin's ability to be a success. Darwin watched two operations that haunted him. After the first year, Darwin's brother left Edinburgh, and Darwin got to know several fellow students interested in natural science, and whom he describes and assesses. For instance, he speaks of Robert Edmund Grant in this context:

> He one day, when we were walking together burst forth in high admiration of Lamarck and his views on evolution. I listened in silent astonishment, and as far as I can judge, without any effect on my mind. I had previously read the *Zoönomia* of my grandfather, in which similar views are maintained, but without producing any effect on me. Nevertheless it is probable that the hearing rather early in life such views maintained and praised may have favoured my upholding them under a different form in my *Origin of Species*. At this time I admired greatly the *Zoönomia*; but on reading it a second time after an interval of ten or fifteen years, I was much disappointed, the proportion of speculation being so large to the facts given.[32]

On evolution, Darwin admits no influence from Lamarck or from his grandfather, Erasmus Darwin, whose work he had admired but later thought it was too given to speculation. Darwin sees himself as coming to uphold evolution in a different form.

Darwin says many things of interest, and, for the sake of space, I will be selective in my discussion. He says that when his father learned that Darwin did not wish to be a doctor, he suggested that his son become a clergyman. Darwin observes:

> Considering how fiercely I have been attacked by the orthodox it seems ludicrous that I once intended to be a clergyman. Nor was this intention and my father's wish ever formally given up, but died a natural death when on leaving Cambridge I joined the *Beagle* as Naturalist.[33]

To become a clergyman, Darwin had to attend an English university, and he had to study classics after he had forgotten much of his Greek, so when he prepared for Cambridge, he had to study and delay his entrance there. His Greek soon recovered, and he translated Homer and the Greek Testament. Darwin was not too enamoured of his education: "During the three years which I spent at Cambridge my time was wasted, as far as the academical studies were concerned, as completely as at Edinburgh and at school."[34] He describes some of his experiences at Cambridge and then gives a sense of the important final year:

> Again in my last year I worked with some earnestness for my final degree of B.A., and brushed up my Classics together with a little Algebra and Euclid, which latter gave me much pleasure, as it did whilst at school. In order to pass the B.A. examination, it was, also, necessary to get up Paley's *Evidences of Christianity*, and his *Moral Philosophy*. This was done in a thorough manner, and I am convinced that I could have written out the whole of the *Evidences* with perfect correctness, but not of course in the clear language of Paley. The logic of this book and as I may add of his *Natural Theology* gave me as much delight as did Euclid. The careful study of these works, without attempting to learn any part by rote, was the only part of the Academical Course which, as I then felt and as I still believe, was of the least use to me in the education of my mind. I did not at that time trouble myself about Paley's premises; and taking these on trust I was charmed and convinced by the long line of argumentation. By answering well the examination questions in Paley, by doing Euclid well, and by not failing miserably in Classics, I gained a good place among the οἱ πολλοί, or crowd of men who do not go in for honours.[35]

Euclid on geometry and Paley on theology were of some use to Darwin and helped him earn his degree at Cambridge. Otherwise, the rest of his studies seemed useless to him. He says the lectures at Edinburgh sickened him, so he skipped them at Cambridge and missed those by Adam Sedgwick, which delayed Darwin becoming a geologist earlier. He did attend John Stevens Henslow's lectures on botany, and although Darwin did not become a botanist, he did go on "delightful" excursions led by Henslow, who, as Nora Barslow notes, helped Darwin get the post on the *Beagle*.[36] Darwin continues to show himself as anything but an ideal person or student and tells stories for and against himself:

> Although as we shall presently see there were some redeeming features in my life at Cambridge, my time was sadly wasted there and worse than wasted. From my passion for shooting and for hunting and when this failed, for riding across country I got into a sporting set, including some dissipated low-minded young men.[37]

Darwin wasted time, and time wasted him to paraphrase and adapt Shakespeare's Richard II.

However, Darwin does salvage some things from his time at Cambridge despite saying that these years were useless. He does speak about his pleasure in art, for instance, at the Fitzwilliam, the National Gallery and a book by Joshua Reynolds, and his enjoyment of music, especially through the choir at King's College chapel. Darwin says that he did not have an ear for music. He adds: "But no pursuit at Cambridge was followed with nearly so much eagerness or gave me so much pleasure as collecting beetles."[38] Darwin has portrayed himself from youth as a collector. He elaborates: "It was the mere passion for collecting, for I did not dissect them and rarely compared their external characters with published descriptions, but got them named anyhow."[39] Part of Darwin's description of his interests in insects while at Cambridge highlights an analogy between poet and entomologist:

> No poet ever felt more delight at seeing his first poem published than I did at seeing in Stephen's *Illustrations of British Insects* the magic words, 'captured by C. Darwin, Esq.' I was introduced to entomology by my second cousin, W. Darwin Fox, a clever and most pleasant man, who was then at Christ's College, and with whom I became extremely intimate.[40]

Darwin expresses his delight, greater than any poet seeing his first poem in print, in seeing the attribution to Darwin in "the magic words." Darwin reports that his cousin introduced him to entomology. Once more, family affects Darwin's mental development and scientific interests. He tries to inject some humour into his narrative: "It seems therefore that a taste for collecting beetles is some indication of future success in life!" but it is also said with some seriousness as Darwin relates this interest to his memory:

> I am surprised what an indelible impression many of the beetles which I caught at Cambridge have left on my mind. I can remember the exact appearance of certain posts, old trees and banks where I made a good capture.[41]

Henslow was an expert in many fields, including entomology. His portrait of Henslow is not only warm and complementary but also combines that with frankness. For instance, he says: "I have not as yet mentioned a circumstance which influenced my whole career more than any other. This was my friendship with Prof. Henslow" while also saying:

> Before long I became well acquainted with Henslow, and during the latter half of my time at Cambridge took long walks with him on most days; so that I was called by some of the dons 'the man who walks with Henslow'.[42]

Darwin humanizes Henslow and himself in his description of this friendship, but he can also be cool or even cold in his assessment. Darwin's prose or rhetoric can be epideictic –combining praise and blame:

> His knowledge was great in botany, entomology, chemistry, mineralogy, and geology. His strongest taste was to draw conclusions from long-continued minute observations. His judgment was excellent, and his whole mind well-balanced; but I do not suppose that anyone would say that he possessed much original genius.[43]

Originality and genius seem to be the ultimate measure for Darwin and although he admits the influence and friendship of Henslow, he does not rate him a genius. The idea of being a genius has classical origins – as in Socrates' daemon – and came to be considered among some in the Renaissance, among the neo-classicists and the Romantics before taking on a more genetic aspect with Darwin's cousin Francis Galton. After talking about how religious and orthodox Henslow was, Darwin gives an example of what Henslow taught him:

> Whilst examining some pollen-grains on a damp surface I saw the tubes exserted, and instantly rushed off to communicate my surprising discovery to him. Now I do not suppose any other Professor of Botany could have helped laughing at my coming in such a hurry to make such a communication. But he agreed how interesting the phenomenon was,

> and explained its meaning, but made me clearly understand how well it was known; so I left him not in the least mortified, but well pleased at having discovered for myself so remarkable a fact, but determined not to be in such a hurry again to communicate my discoveries.[44]

Darwin includes biographical sketches of others in his autobiography, and he has a good sense of story (perhaps not unlike his father) and detail. Henslow, although not a genius, taught Darwin something. All this according to Darwin. Among these portraits, Henslow's is the most detailed. Darwin uses these sketches to return to his self-portrait:

> Looking back, I infer that there must have been something in me a little superior to the common run of youths, otherwise the above-mentioned men, so much older than me and higher in academical position, would never have allowed me to associate with them. Certainly I was not aware of any such superiority, and I remember one of my sporting friends, Turner, who saw me at work on my beetles, saying that I should some day be a Fellow of the Royal Society, and the notion seemed to me preposterous.[45]

Darwin looks back and sees a superiority in himself he did not see himself, and he shows this through an inference looking through the eyes of others, these high academics and even his sporting friend. The "preposterous" became less so, and Darwin did become a fellow of the Royal Society and one of its most celebrated fellows. He was and is modest, but that modesty can also recognize his accomplishments and, by implication, his genius. Darwin seems to combine frankness and humility.

Reading affected Darwin, as he recounts, and he has said that he preferred reading books to attending lectures:

> During my last year at Cambridge I read with care and profound interest Humboldt's *Personal Narrative*. This work and Sir J. Herschel's *Introduction to the Study of Natural Philosophy* stirred up in me a burning zeal to add even the most humble contribution to the noble structure of Natural Science. No one or a dozen other books influenced me nearly so much as these two.[46]

Despite Darwin's earlier claim that he did not learn much formally at Cambridge, he admits to learning from Henslow and from his reading and has fond memories of the university:

> Upon the whole the three years which I spent at Cambridge were the most joyful in my happy life; for I was then in excellent health, and almost always in high spirits.

> As I had at first come up to Cambridge at Christmas, I was forced to keep two terms after passing my final examination, at the commencement of 1831; and Henslow then persuaded me to begin the study of geology.[47]

Happiness and the influence of Henslow, this time in advising Darwin to study geology, are an important part of Darwin's narrative. He is happy to attribute influence to Henslow, but he is less apt to see affinities between his work on evolution and that of Lamarck and of Erasmus Darwin, his grandfather. Whether Darwin is being precise or suffering from the burden of the past or the anxiety of influence with those closest to his own field is an open question.[48] He also mentions more of those who influenced him and provided him with opportunities: "Professor Sedgwick intended to visit N. Wales in the beginning of August to pursue his famous geological investigation amongst the older rocks, and Henslow asked him to allow me to accompany him."[49] He also mentions Sedgwick in connection with Henslow and in the context of geology. He relates Sedgwick to science:

> But I was then utterly astonished at Sedgwick not being delighted at so wonderful a fact as a tropical shell being found near the surface in the middle of England. Nothing before had ever made me thoroughly realise, though I had read various scientific books, that science consists in grouping facts so that general laws or conclusions may be drawn from them.[50]

Here, from his reading and experience, Darwin defines science.

The world beyond Cambridge awaited Darwin. The next chapter begins with this voyage and the change it made to Darwin and the world of science. The perception of life and time on earth would never be the same.

Notes

1 Francis Darwin and Nora Barlow, the son and granddaughter of Darwin, edit and write about the life and work of Charles Darwin. For related work on Charles Darwin, see, for instance, Gillian Beer, *Darwin's Plots : Evolutionary Narrative in Darwin, George Eliot, and Nineteenth-Century Fiction* (London : Routledge & Kegan Paul, 1983); Robert M. Young, *Darwin's Metaphor: Nature's Place in Victorian Culture* (Cambridge: Cambridge University Press, 1985); Janet Browne, *Charles Darwin: Voyaging* (New York: Knopf, 1995); Janet Browne, *Charles Darwin: The Power of Place* (Princeton, NJ: Princeton University Press, 2002); Paul Johnson, *Darwin: Portrait of a Genius* (New York: Viking, 2012); Andrew Norman Wilson, *Charles Darwin: Victorian Mythmaker* (London: John Murray, 2017); Michael Ruse, *Charles Darwin: No Rebel, Great Revolutionary* (Cambridge: Cambridge University Press, 2024).

2 Francis Darwin, ed. *The Life and Letters of Charles Darwin, Including an Autobiographical Chapter* (London: John Murray, 1887), Volume 1, page 26.

3 Nora Barlow, "Preface," in *The Autobiography of Charles Darwin 1809–1882 with Original Omissions Restored*, edited with Appendix and Notes by his grand-daughter Nora Barlow (London: Collins, 1958), 5.
4 Barlow, "Preface," 6.
5 Darwin, *Autobiography*, ed. Nora Barlow, 21.
6 Darwin, *Autobiography*, 21–22.
7 Darwin, *Autobiography*, 22–23.
8 Darwin, *Autobiography*, 23.
9 Darwin, *Autobiography*, 27.
10 Darwin, *Autobiography*, 27.
11 Darwin, *Autobiography*, 27.
12 Darwin, *Autobiography*, 28.
13 Darwin, *Autobiography*, 28.
14 Darwin, *Autobiography*, 28.
15 Darwin, *Autobiography*, 29.
16 Darwin, *Autobiography*, 29–30.
17 Darwin, *Autobiography*, 32.
18 Darwin, *Autobiography*, 36.
19 Darwin, *Autobiography*, 39.
20 Darwin, *Autobiography*, 42.
21 Darwin, *Autobiography*, 42.
22 Darwin, *Autobiography*, 43. On Galton, see Dominique Aubert-Marson, "Sir Francis Galton : le fondateur de l'eugénisme," *Medical Science* (Paris) 25 (6–7) (2009): 641–645.
23 R. E. Fancher, "Scientific Cousins: The Relationship between Charles Darwin and Francis Galton," *American Psychology*, February-March, 64 (2) (2009): 84–92.
24 Darwin, *Autobiography*, 43.
25 Darwin, *Autobiography*, 43–44.
26 Darwin, *Autobiography*, 44.
27 Evelyn M. Witkin, "The Wonders of Darwin's World," *Perspectives in Biology and Medicine* 40 (4) (1997): 471–478.
28 Darwin, *Autobiography*, 44.
29 Darwin, *Autobiography*, 45–46.
30 Darwin, *Autobiography*, 46.
31 Darwin, *Autobiography*, 46.
32 Darwin, *Autobiography*, 49.
33 Darwin, *Autobiography*, 57.
34 Darwin, *Autobiography*, 58.
35 Darwin, *Autobiography*, 58–59.
36 Darwin, *Autobiography*, 60, see 59.
37 Darwin, *Autobiography*, 60.
38 Darwin, *Autobiography*, 62.
39 Darwin, *Autobiography*, 62.
40 Darwin, *Autobiography*, 63.
41 Darwin, *Autobiography*, 63.
42 Darwin, *Autobiography*, 64.
43 Darwin, *Autobiography*, 64.
44 Darwin, *Autobiography*, 66.
45 Darwin, *Autobiography*, 67.
46 Darwin, *Autobiography*, 67–68.

47 Darwin, *Autobiography*, 68.
48 Walter Jackson Bate, *The Burden of the Past and the English Poet* (Cambridge, MA: Belknap Press, 1970); Harold Bloom, *The Anxiety of Influence: A Theory of Poetry* (New York: Oxford University Press, 1973).
49 Darwin, *Autobiography*, 69.
50 Darwin, *Autobiography*, 69–70.

2 Charles Darwin

From the Beagle to *Origin of the Species*

Darwin's self-representation shows that this was a key time of his life. The next phase of Darwin's life begins:

> ON RETURNING home from my short geological tour in N. Wales, I found a letter from Henslow, informing me that Captain Fitz-Roy was willing to give up part of his own cabin to any young man who would volunteer to go with him without pay as naturalist to the Voyage of the *Beagle*.[1]

This opportunity, as Darwin mentioned earlier, was a key to his life, his experience and science. Darwin says his father objected but with the fortunate caveat – "If you can find any man of common sense, who advises you to go, I will give my consent" – so after Darwin refused the offer, his uncle, who thought the voyage a good idea and whom Darwin's father thought "one of the most sensible men in the world," got that consent for Darwin to sail on the ship.[2]

Many telling stories and details occur in Darwin's *Autobiography*, so I can only be selective. Darwin's portrait of Captain Fitz-Roy of the *Beagle* involves his usual sense of pros and cons:

> He seemed to me often to fail in sound judgment or common sense. He was extremely kind to me, but was a man very difficult to live with on the intimate terms which necessarily followed from our messing by ourselves in the same cabin. We had several quarrels; for when out of temper he was utterly unreasonable. For instance, early in the voyage at Bahia in Brazil he defended and praised slavery, which I abominated, and told me that he had just visited a great slave-owner, who had called up many of his slaves and asked them whether they were happy, and whether they wished to be free, and all answered 'No.' I then asked him, perhaps with a sneer, whether he thought that the answers of slaves in the presence of their master was worth anything. This made him excessively

DOI: 10.4324/9781003651871-3

> angry, and he said that as I doubted his word, we could not live any longer together. I thought that I should have been compelled to leave the ship; but as soon as the news spread, which it did quickly, as the captain sent for the first lieutenant to assuage his anger by abusing me, I was deeply gratified by receiving an invitation from all the gun-room officers to mess with them. But after a few hours Fitz-Roy showed his usual magnanimity by sending an officer to me with an apology and a request that I would continue to live with him.[3]

I quote at length because this passage shows this almost oxymoronic portrait that often characterizes Darwin's narrative in this work. Here, Fitz-Roy is "extremely kind" but "difficult" to live with in the same cabin. Darwin and he quarrelled, for instance over slavery, which Fitz-Roy praised while Darwin abominated it. For Darwin, slaves paraded and denying they wanted to be free was worthless testimony. Darwin uses intensive and contrasting diction to underscore his point, Darwin sneered and that made Fitz-Roy angry – Darwin uses the polyptoton of "angry" and "anger" for emphasis. A few hours later, Fitz-Roy, through an officer, apologizes and requests that Darwin keep living with him. Darwin dramatizes opposition, conflict and reversals and gives his autobiography some sense of the human and narrative energy.

The extended portrait of Fitz-Roy continues this bringing together of ups and downs, this collision of opposites and this division and reconciliation:

> I saw Fitz-Roy only occasionally after our return home, for I was always afraid of unintentionally offending him, and did so once, almost beyond mutual reconciliation. He was afterwards very indignant with me for having published so unorthodox a book (for he became very religious) as the *Origin of Species*. Towards the close of his life he was as I fear, much impoverished, and this was largely due to his generosity. Anyhow after his death a subscription was raised to pay his debts. His end was a melancholy one, namely suicide, exactly like that of his uncle Ld. Castlereagh, whom he resembled closely in manner and appearance.
>
> His character was in several respects one of the most noble which I have ever known, though tarnished by grave blemishes.[4]

Darwin tells a good story. He understands conflict as one of the elements of drama and the dramatic nature of narrative. Small details matter. The phrase "mutual reconciliation" focuses on the relation and seems to imply that while Darwin was willing to reconcile, Fitz-Roy was not. The return home leads to offence without reconciliation as well, in time, with Fitz-Roy's new-found religious feelings, indignation over Darwin's *Origin of the Species*. Darwin looks back wistfully at the melancholy end of

Fitz-Roy, a suicide like his famous uncle. Darwin reveals his own character while representing others such as the Captain – most noble but blemished.

For Darwin, there is one central matter in his life, and he makes sure that he continues to stress the voyage:

> The voyage of the *Beagle* has been by far the most important event in my life and has determined my whole career; yet it depended on so small a circumstance as my uncle offering to drive me 30 miles to Shrewsbury, which few uncles would have done, and on such a trifle as the shape of my nose. I have always felt that I owe to the voyage the first real training or education of my mind. I was led to attend closely to several branches of natural history, and thus my powers of observation were improved, though they were already fairly developed.[5]

Darwin repeats the point he made earlier in his autobiography. The reiteration includes making the joining of the voyage a very human matter, giving his uncle credit and telling the tale of others judging Darwin by his nose, the claim that the voyage was his real education. Darwin also expresses his usual qualification of his own statements, so that even though he learned more about natural history and how to observe, he already had fairly powerful observation. Summary and repetition are part of his technique, a kind of reiteration with a difference. The effect is cumulative.

Reading continues to be important to Darwin and a theme he stresses in his autobiography:

> The investigation of the geology of all the places visited was far more important, as reasoning here comes into play. On first examining a new district nothing can appear more hopeless than the chaos of rocks; but by recording the stratification and nature of the rocks and fossils at many points, always reasoning and predicting what will be found elsewhere, light soon begins to dawn on the district, and the structure of the whole becomes more or less intelligible. I had brought with me the first volume of Lyell's *Principles of Geology*, which I studied attentively; and this book was of the highest service to me in many ways. The very first place which I examined, namely St. Jago in the Cape Verde islands, showed me clearly the wonderful superiority of Lyell's manner of treating geology, compared with that of any other author, whose works I had with me or ever afterwards read.[6]

Darwin's interest in geology is a key to his scientific work, and this passage builds on his earlier descriptions of his geological work. Charles Lyell was a great influence on Darwin, as he admits, and later became a friend. Lyell, born about twelve years before Darwin and someone who faced opposition in his views of geology, wrote to Darwin in 1837:

> I could think of nothing for days after your lesson on coral reefs, but of the tops of submerged continents.
>
> It is all true, but do not flatter yourself that you will be believed, till you are growing bald, like me with hard work, & vexation at the incredulity of the world.[7]

It is hard to change views in any human endeavour, including science. When I was an undergraduate studying the history of science, Thomas Kuhn's work on the structure of scientific revolutions was still much in evidence.[8] That included the reluctance for scientists to embrace a new paradigm. Lyell himself wants Darwin to know about the incredulity.

At different times in this autobiography, Darwin reflects on science and on his role in it. Here, he comes to a series of reflections:

> Another of my occupations was collecting animals of all classes, briefly describing and roughly dissecting many of the marine ones; but from not being able to draw and from not having sufficient anatomical knowledge a great pile of MS. which I made during the voyage has proved almost useless. I thus lost much time, with the exception of that spent in acquiring some knowledge of the Crustaceans, as this was of service when in after years I undertook a monograph of the Cirripedia.[9]

Darwin describes his activities of the collection of animals, his lack of knowledge of drawing and anatomy and the uselessness of this work and his loss of time, except in learning about crustaceans. He is frank not only with others but also with himself. Darwin weighs evidence even regarding himself and his own method, his successes and failures. He also speaks about his writing:

> During some part of the day I wrote my Journal, and took much pains in describing carefully and vividly all that I had seen; and this was good practice. My Journal served, also, in part as letters to my home, and portions were sent to England, whenever there was an opportunity.[10]

The writing helped observation and served as communication with his family. Darwin is a scientist and writer. Once more, he stresses the importance of the expedition:

> The above various special studies were, however, of no importance compared with the habit of energetic industry and of concentrated attention

> to whatever I was engaged in, which I then acquired. Everything about which I thought or read was made to bear directly on what I had seen and was likely to see; and this habit of mind was continued during the five years of the voyage. I feel sure that it was this training which has enabled me to do whatever I have done in science.[11]

Darwin extols concentration and says that, for the five years of the voyage, all his thoughts and reading bore on what he saw or was to see. He views this as training and considers it to have enabled him to achieve what he did in science. Here is Darwin's method, his training away from a formal school or university. His retrospection leads him to consider science:

> Looking backwards, I can now perceive how my love for science gradually preponderated over every other taste. During the first two years my old passion for shooting survived in nearly full force, and I shot myself all the birds and animals for my collection; but gradually I gave up my gun more and more, and finally altogether to my servant, as shooting interfered with my work, more especially with making out the geological structure of a country. I discovered, though unconsciously and insensibly, that the pleasure of observing and reasoning was a much higher one than that of skill and sport.[12]

Darwin glosses this description with further interpretation. He shows the movement of his life, important changes:

> The primeval instincts of the barbarian slowly yielded to the acquired tastes of the civilized man. That my mind became developed through my pursuits during the voyage, is rendered probable by a remark made by my father, who was the most acute observer whom I ever saw, of a sceptical disposition, and far from being a believer in phrenology; for on first seeing me after the voyage, he turned round to my sisters and exclaimed, 'Why, the shape of his head is quite altered.'[13]

Darwin represents his own behaviour as moving from barbarity to civilization and brings his father, the acute observer and sceptic, back into his account, referring to his father's disbelief in phrenology. Darwin portrays himself implicitly as a man of good humour and spirits. Darwin represents character to give his prose character.

For the autobiography, Darwin gives a sense of what it is and is not doing. For instance, he writes: "I need not here refer to the events of the voyage—where we went and what we did—as I have given a sufficiently full account in my published Journal."[14] He also assesses his own contribution to science on the journey:

> I also reflect with high satisfaction on some of my scientific work, such as solving the problem of coral-islands, and making out the geological structure of certain islands, for instance, St. Helena. Nor must I pass over the discovery of the singular relations of the animals and plants inhabiting the several islands of the Galapagos archipelago, and of all of them to the inhabitants of South America.[15]

This summary stresses his satisfaction with his science, his contribution to geology and his discovery on the Galapagos archipelago. He also looks for his motivation:

> As far as I can judge of myself I worked to the utmost during the voyage from the mere pleasure of investigation, and from my strong desire to add a few facts to the great mass of facts in natural science. But I was also ambitious to take a fair place among scientific men,—whether more ambitious or less so than most of my fellow-workers I can form no opinion.[16]

He portrays his motive as one of pleasure that investigation brought and he admits his own ambition as a scientist and cannot opine whether he was more or less ambitious than those with whom he worked. Darwin expresses his ambition but qualifies it. He does not have the evidence to put it in context. Responding to others or comparing himself to others helps to define Darwin. He listens to others in his circle whom he respects: "Fitz-Roy asked to read some of my Journal, and declared it would be worth publishing; so here was a second book in prospect!"[17] Darwin is precise about what he thinks the views others had of him meant to him:

> Towards the close of our voyage I received a letter whilst at Ascension, in which my sisters told me that Sedgwick had called on my father and said that I should take a place among the leading scientific men. I could not at the time understand how he could have learnt anything of my proceedings, but I heard (I believe afterwards) that Henslow had read some of the letters which I wrote to him before the Philosophical Soc. of Cambridge and had printed them for private distribution. My collection of fossil bones, which had been sent to Henslow, also excited considerable attention amongst palæontologists. After reading this letter I clambered over the mountains of Ascension with a bounding step and made the volcanic rocks resound under my geological hammer! All this shows

> how ambitious I was; but I think that I can say with truth that in after years, though I cared in the highest degree for the approbation of such men as Lyell and Hooker, who were my friends, I did not care much about the general public. I do not mean to say that a favourable review or a large sale of my books did not please me greatly; but the pleasure was a fleeting one, and I am sure that I have never turned one inch out of my course to gain fame.[18]

Darwin reports what he was told what Sedgwick told Darwin's father about Darwin being a leading scientist and how Henslow did to get word out about Darwin's work and the effect this had on Darwin and how it showed his ambition. Darwin says that he cared about the approbation of his friends like Lyell and Hooker but not much about that of the public, even if he had a fleeting pleasure if his books were well reviewed or sold well. Fame, he says, never took him off course. Letters to Darwin and by him played a key role during his voyage. The epistles figure in his autobiography. He was a man of letters.

After his return from the voyage, Darwin spent time in Cambridge. His writing and his friends continue to play a role in this autobiography. He notes that "I settled in lodgings at Cambridge on December 13th, where all my collections were under the care of Henslow. I stayed here three months and got my minerals and rocks examined by the aid of Prof. Miller" and then says that

> I began preparing my Journal of travels, which was not hard work, as my MS. Journal had been written with care, and my chief labour was making an abstract of my more interesting scientific results. I sent also, at the request of Lyell, a short account of my observations on the elevation of the coast of Chile to the Geological Society.[19]

Darwin emphasizes the care he took in his writing. Henslow and Lyell are instrumental in supporting Darwin, which he amplifies here. Darwin highlights Lyell and discusses his own meetings with Robert Brown, a botanist. Darwin details the work he did in London and his esteem for Lyell:

> On March 7th, 1837, I took lodgings in Great Marlborough Street in London and remained there for nearly two years until I was married. During these two years I finished my *Journal*, read several papers before the Geological Society, began preparing the MS. for my *Geological Observations* and arranged for the publication of the *Zoology of the Voyage of the Beagle*. In July I opened my first note-book for facts in relation to the *Origin of Species*, about which I had long reflected, and never ceased working on for the next twenty years.

> During these two years I also went a little into society, and acted as one of the hon. secretaries of the Geological Society. I saw a great deal of Lyell. One of his chief characteristics was his sympathy with the work of others, ...[20]

Darwin stresses his long and patient labours in geology and zoology and in his publications. Reflection and writing are central activities that Darwin represents. He speaks about one of his own papers that was a failure, which he seems to have realized after Agassiz's glacier-lake theory. As Darwin says, being unable to work on science all day, he read books not only on metaphysics, for which he was not fitted, but also of poetry:

> I took much delight in Wordsworth's and Coleridge's poetry, and can boast that I read the *Excursion* twice through. Formerly Milton's *Paradise Lost* had been my chief favourite, and in my excursions during the voyage of the *Beagle*, when I could take only a single small volume, I always chose Milton.[21]

Whereas once Darwin read Milton on the voyage of the *Beagle* (1831–1836), he is now reading William Wordsworth and Samuel Taylor Coleridge. Coleridge died in 1834 while Darwin was on the voyage while Wordsworth lived to 1850. These Romantic poets, both of them extraordinary, have a keen interest in nature and philosophical observation in their poetry.

Darwin then shifts to religion:

> DURING THESE two years I was led to think much about religion. Whilst on board the Beagle I was quite orthodox, and I remember being heartily laughed at by several of the officers (though themselves orthodox) for quoting the Bible as an unanswerable authority on some point of morality.[22]

On the voyage, Darwin portrays himself as someone orthodox who recognized the Bible as the ultimate authority. Here, he also chronicles a change:

> But I had gradually come, by this time, to see that the Old Testament from its manifestly false history of the world, with the Tower of Babel, the rainbow as a sign, etc., etc., and from its attributing to God the feelings of a revengeful tyrant, was no more to be trusted than the sacred books of the Hindoos, or the beliefs of any barbarian. The question then continually rose before my mind and would not be banished,—is it credible that if God were now to make a revelation to the Hindoos, would he permit it to be connected with the belief in Vishnu, Siva, &c., as Christianity is connected with the Old Testament. This appeared to me utterly incredible.[23]

Darwin sees the Old Testament as "false" and he uses the polyptoton of "credible" and "incredible" to show that he does see this part of the Bible as an untrue historical account of the world. He speaks of God and compares Christians and Hindus in terms of revelation. The adverbs also intensify Darwin's point as in "manifestly false" and "utterly incredible" while also using an adjective in a phrase describing the "attributing to God the feelings of a revengeful tyrant." Darwin shows how he moved from orthodoxy to not believing at least the incredible and false version of history in the Old Testament. Darwin gives evidence of why he moved away from his earlier orthodox beliefs, but rather than go into those details, I quote his conclusion: "I gradually came to disbelieve in Christianity as a divine revelation."[24] If there is any doubt of where Darwin stands, he soon adds: "Beautiful as is the morality of the New Testament, it can hardly be denied that its perfection depends in part on the interpretation which we now put on metaphors and allegories."[25] The literal leads to the figurative and that depends on interpreting. He was not willing to give up his belief and hoped for evidence to believe, but gradually this all changed:

> Thus disbelief crept over me at a very slow rate, but was at last complete. The rate was so slow that I felt no distress, and have never since doubted even for a single second that my conclusion was correct. I can indeed hardly see how anyone ought to wish Christianity to be true; for if so the plain language of the text seems to show that the men who do not believe, and this would include my Father, Brother and almost all my best friends, will be everlastingly punished.
>
> And this is a damnable doctrine.[26]

According to Nora Barlow, six months after the death of Darwin, his wife annotated this passage after "distress" – a comment that few then would call Christianity the doctrine of everlasting damnation for disbelief – and this part of Darwin's narrative was not included in the original.[27] Her annotation seems to be an attempt to reconcile Darwin with Christianity. Barlow includes the whole passage. Darwin's disbelief comes slowly and completely and its correctness he never doubted thereafter. He thinks this doctrine to be damnable for damning those closest to Darwin. For the reader, Darwin leaves no doubt about his movement from belief to disbelief. Nor is Darwin convinced by intelligent design:

> The old argument of design in nature, as given by Paley, which formerly seemed to me so conclusive, fails, now that the law of natural selection has been discovered. We can no longer argue that, for instance, the beautiful hinge of a bivalve shell must have been made by an intelligent being, like the hinge of a door by man. There seems to be no more

> design in the variability of organic beings and in the action of natural selection, than in the course which the wind blows. Everything in nature is the result of fixed laws.[28]

Natural selection belies design in nature, which fixed laws govern. Darwin is not reconciling Christianity and science, but this aspect of culture and science. He has much to say about religion including his doubts and his father's advice:

> Nothing is more remarkable than the spread of scepticism or rationalism during the latter half of my life. Before I was engaged to be married, my father advised me to conceal carefully my doubts, for he said that he had known extreme misery thus caused with married persons.[29]

Religious views can affect marriage, family and friendship. Darwin's religious beliefs were a delicate matter for his family, including his wife, whom he praises:

> I marvel at my good fortune that she, so infinitely my superior in every single moral quality, consented to be my wife. She has been my wise adviser and cheerful comforter throughout life, which without her would have been during a very long period a miserable one from ill-health.[30]

Religion was not a topic that Darwin could choose to avoid.

Among Darwin's further comments on science are those about Lyell. Descriptions of people in the autobiography can be splintered or what I once called long ago now the Osiris effect in narrative:

> I saw more of Lyell than of any other man both before and after my marriage. His mind was characterised, as it appeared to me, by clearness, caution, sound judgment and a good deal of originality. When I made any remark to him on Geology, he never rested until he saw the whole case clearly and often made me see it more clearly than I had done before. He would advance all possible objections to my suggestion, and even after these were exhausted would long remain dubious. A second characteristic was his hearty sympathy with the work of other scientific men.[31]

Darwin sees originality in Lyell, something unusual in his estimation. Lyell is also good at testing ideas and, as Darwin said before, he was interested in the work of other scientists. As is Darwin's wont, he weighs Lyell's strengths and weaknesses: One weakness was his overestimation of rank.[32] He remembers Henslow recommending Lyell's work to Darwin and praises Lyell's accomplishment: "The powerful effects of Lyell's works could formerly be plainly seen in the different progress of the science in France and England."[33] Darwin considers other geologists such as William Buckland and Roderick Impey Murchison and does so in his usual ambivalent manner, finding Buckland vulgar and Murchison too impressed by rank. Darwin returns to a description of Robert Brown who was, in Darwin's view, both miserly and generous.[34] According to Darwin, Richard Owen, whom Darwin admired, was jealous of the success of *Origin of the Species*; Joseph Dalton Hooker, a good friend, had an acute intellect but could be indignant over small details; and Huxley, the best talker Darwin says he had known, had a sharp mind and was able to expose the errors of other scientists, which he thought was an important function.[35] Darwin has something to say about others he met, such as John Frederick William Herschel, Humboldt, Charles Babbage, Herbert Spencer, Henry Thomas Buckle, Sydney Smith, Macaulay, Carlyle and others. Darwin weighs the positive and negative. For instance, Darwin says about Buckle's *History of Civilisation*: "This book I thought most interesting and read it twice; but I doubt whether his generalisations are worth anything."[36] Darwin often considers the relation between induction and deduction, the specific and the general. He applies his scientific way of thinking and his sense of logic to scientific and cultural texts. This is something I also observe in the writing of Bertrand Russell. Of Thomas Carlyle, Darwin says: "He has been all-powerful in impressing some grand moral truths on the minds of men. On the other hand, his views about slavery were revolting."[37] At another moment, Darwin shows an example of Carlyle's sneering:

> It is astonishing to me that Kingsley should have spoken of him as a man well fitted to advance science. He laughed to scorn the idea that a mathematician, such as Whewell, could judge, as I maintained he could, of Goethe's views on light.[38]

Darwin is not impressed by Carlyle's lack of science and his scorn for scientists. Darwin is categorical in his conclusion: "As far as I could judge, I never met a man with a mind so ill adapted for scientific research."[39] It seems that Darwin sees Carlyle as seeing culture and science as being distinct. Light is something of interest to scientists. Darwin is interested in scientists and science and all this being out in society might have been too much for Darwin and his health, and his wife and he were happy to move to the country.[40] Darwin's exhaustion and ill health, he says, made it hard to see anyone, except his wife and children, for more than an hour and this situation affected his autobiography:

> My chief enjoyment and sole employment throughout life has been scientific work; and the excitement from such work makes me for the time

> forget, or drives quite away, my daily discomfort. I have therefore nothing to record during the rest of my life, except the publication of my several books.[41]

Darwin needed the quiet life to attend to his health.

Darwin sets out the sales and interest in his publications. Amidst the discussion of these books and their contribution, he tells that in late 1847 his health was so bad that he could not attend the funeral or act as executor of his father. He also shifts to a later time: "From September 1854 onwards I devoted all my time to arranging my huge pile of notes, to observing, and experimenting, in relation to the transmutation of species."[42] Darwin then goes back in time. He returns to Lyell's influence on his method:

> After my return to England it appeared to me that by following the example of Lyell in Geology, and by collecting all facts which bore in any way on the variation of animals and plants under domestication and nature, some light might perhaps be thrown on the whole subject.[43]

Darwin is more specific of the timing and activity:

> My first note-book was opened in July 1837. I worked on true Baconian principles, and without any theory collected facts on a wholesale scale, more especially with respect to domesticated productions, by printed enquiries, by conversation with skilful breeders and gardeners, and by extensive reading.[44]

Darwin expresses his retrospective surprise at his own industry. He shows how his ideas had been developed in thought and manuscript long before they appeared. He gives an instance of someone else, Alfred Russel Wallace, coming to the same conclusions as he did.

> Early in 1856 Lyell advised me to write out my views pretty fully, and I began at once to do so on a scale three or four times as extensive as that which was afterwards followed in my *Origin of Species*; yet it was only an abstract of the materials which I had collected, and I got through about half the work on this scale. But my plans were overthrown, for early in the summer of 1858 Mr Wallace, who was then in the Malay archipelago, sent me an essay *On the Tendency of Varieties to Depart Indefinitely from the Original Type*; and this essay contained exactly the same theory as mine. Mr Wallace expressed the wish that if I thought well of his essay, I should send it to Lyell for perusal.[45]

There was, then, a challenge that Wallace wrote Darwin and had come to the same views as Darwin, so how should Darwin handle this problem? Originality and priority are important to Darwin and in science. He elaborates:

> The circumstances under which I consented at the request of Lyell and Hooker to allow of an extract from my MS., together with a letter to Asa Gray, dated September 5, 1857, to be published at the same time with Wallace's Essay, are given in the *Journal of the Proceedings of the Linnean Society*, 1858, p. 45. I was at first very unwilling to consent, as I thought Mr Wallace might consider my doing so unjustifiable, for I did not then know how generous and noble was his disposition. The extract from my MS. and the letter to Asa Gray had neither been intended for publication, and were badly written. Mr Wallace's essay, on the other hand, was admirably expressed and quite clear. Nevertheless, our joint productions excited very little attention, and the only published notice of them which I can remember was by Professor Haughton of Dublin, whose verdict was that all that was new in them was false, and what was true was old. This shows how necessary it is that any new view should be explained at considerable length in order to arouse public attention.[46]

Darwin draws conclusions about how to express and deliver scientific discoveries. Darwin worked in a way that included conversations, manuscripts, notebooks and letters, like the one to Asa Gray (a figure I discuss in another chapter) in additions to scientific meetings and publications.

The publication Darwin then stresses is his most important example:

> In September 1858 I set to work by the strong advice of Lyell and Hooker to prepare a volume on the transmutation of species, but was often interrupted by ill-health, and short visits to Dr. Lane's delightful hydropathic establishment at Moor Park. I abstracted the MS. begun on a much larger scale in 1856, and completed the volume on the same reduced scale. It cost me thirteen months and ten days' hard labour. It was published under the title of the *Origin of the Species*, in November 1859. Though considerably added to and corrected in the later editions, it has remained substantially the same book.[47]

Darwin returns to the role of his scientific friends, Lyell and Hooker, and mentions his ill health once more and the nature of his key book. He assesses that contribution:

> It is no doubt the chief work of my life. It was from the first highly successful. The first small edition of 1250 copies was sold on the day

> of publication, and a second edition of 3000 copies soon afterwards. Sixteen thousand copies have now (1876) been sold in England and considering how stiff a book it is, this is a large sale. It has been translated into almost every European tongue, even into such languages as Spanish, Bohemian, Polish, and Russian.[48]

Darwin sees this book as his chief work, and he gauges success partly by sales and translations. He goes on to talk about the translation of the book into Japanese as well as essay in Hebrew, the reviews of all Darwin's works, so numerous that he has stopped collecting them (a hard decision for a natural collector), and the essays, books and bibliography on the subject. The success Darwin elaborates on:

> The success of the *Origin* may, I think, be attributed in large part to my having long before written two condensed sketches, and to my having finally abstracted a much larger manuscript, which was itself an abstract. By this means I was enabled to select the more striking facts and conclusions. I had, also, during many years, followed a golden rule, namely, that whenever a published fact, a new observation or thought came across me, which was opposed to my general results, to make a memorandum of it without fail and at once; for I had found by experience that such facts and thoughts were far more apt to escape from the memory than favourable ones. Owing to this habit, very few objections were raised against my views which I had not at least noticed and attempted to answer.[49]

Darwin distils his assessment of the success of *Origin* to abstraction and selection as well as a memorandum on something that opposed his general results. That way the book could be memorable and address objections before they were raised. Naturalists still resisted because they did not doubt that species were permanent:

> Even Lyell and Hooker, though they would listen with interest to me, never seemed to agree. I tried once or twice to explain to able men what I meant by natural selection, but signally failed. What I believe was strictly true is that innumerable well-observed facts were stored in the minds of naturalists, ready to take their proper places as soon as any theory which would receive them was sufficiently explained. Another element in the success of the book was its moderate size; and this I owe to the appearance of Mr Wallace's essay; had I published on the scale in which I began to write in 1856, the book would have been four or five times as large as the *Origin*, and very few would have had the patience to read it.[50]

It was not easy for Darwin to have his theory accepted. Wallace helped to bring Darwin to a more compact *Origin*, something more amenable to readers. Darwin then goes to a concern he has had throughout his autobiography – the originality of others and of himself:

> I gained much by my delay in publishing from about 1839, when the theory was clearly conceived, to 1859; and I lost nothing by it, for I cared very little whether men attributed most originality to me or Wallace; and his essay no doubt aided in the reception of the theory. I was forestalled in only one important point, which my vanity has always made me regret, namely, the explanation by means of the Glacial period of the presence of the same species of plants and of some few animals on distant mountain summits and in the arctic regions.[51]

He argues that the delay was beneficial and that he did not care whether Wallace or he originated the theory. That view might be contrary to his assessment of other scientists as original or not. This is his chief work, and would he really want it to be original? He does give Wallace credit in helping the theory be received. Darwin portrays his own vanity as the reason the work was stalled. He was vain in his explanation of the glacial period. Nor is Darwin easy to represent as his self-representation is ambivalent in this autobiography. He admits his ambition and hides his ambition. He values originality but then says that it does not matter in his chief book. He then discusses his own work in relation to that of others, those who had worked on similar theories, so if priority and originality were not an issue, why examine his own work in relation to that of others? He mentions the work of E. Forbes, Fritz Müller and Ernst Häckel. It would be a mistake to underestimate the intellectual generosity of Darwin. Soon after Darwin died, Müller wrote a letter to Ernst Krause, saying that he told Darwin his experiences in natural science and that Darwin sent him suggestions and adding: "The wonderful qualities of his character, this total freedom from any petty conceit, this temperateness in judgement, this ungrudging, often too ready acknowledgement of others had to endear him to everyone;…"[52] Darwin has a certain magnanimity to and understanding of his reviewers:

> This leads me to remark that I have almost always been treated honestly by my reviewers, passing over those without scientific knowledge as not worthy of notice. My views have often been grossly misrepresented, bitterly opposed and ridiculed, but this has been generally done, as I believe, in good faith.[53]

Darwin is forgiving but also does not pay attention to those reviewers who do not know science. Once more, Lyell gave Darwin good advice:

> I rejoice that I have avoided controversies, and this I owe to Lyell, who many years ago, in reference to my geological works, strongly advised me never to get entangled in a controversy, as it rarely did any good and caused a miserable loss of time and temper.[54]

Darwin also had his own way of coping with the response to his work:

> Whenever I have found out that I have blundered, or that my work has been imperfect, and when I have been contemptuously criticised, and even when I have been overpraised, so that I have felt mortified, it has been my greatest comfort to say hundreds of times to myself that 'I have worked as hard and as well as I could, and no man can do more than this.' I remember when in Good Success Bay, in Tierra del Fuego, thinking, (and I believe that I wrote home to the effect) that I could not employ my life better than in adding a little to natural science. This I have done to the best of my abilities, and critics may say what they like, but they cannot destroy this conviction.[55]

Darwin focuses on doing one's best and the belief he has done that in natural science and that one cannot allow praise and dispraise to sway one from that conviction. Epideictic rhetoric did not sway Darwin: That is how he represents himself. Darwin is also attentive to the advice of others, as I noted earlier, and he reads what is recommended to him: He reads "through the advice of Robert Brown, a copy of C. K. Sprengel's wonderful book, *Das entdeckte Geheimnis der Natur*" and soon concludes: "The merits of poor old Sprengel, so long overlooked, are now fully recognised many years after his death."[56] Darwin is generous in his praise of those who have made a contribution to science.

Another important phase now occurred in his life. His move to a more quiet place away from urban life allowed him time for peace and reflection. The next chapter will start its analysis at this point.

Notes

1 Charles Darwin, *The Autobiography of Charles Darwin 1809–1882 with Original Omissions Restored*, edited with Appendix and Notes by his granddaughter Nora Barlow (London: Collins, 1958), 71.
2 Darwin, *Autobiography*, 72.
3 Darwin, *Autobiography*, 73–74.

4 Darwin, *Autobiography*, 76.
5 Darwin, *Autobiography*, 76–77.
6 Darwin, *Autobiography*, 77.
7 Charles Lyell, Letter to Charles Darwin, 13 February 1837, Darwin Correspondence Project, "Letter no. 343," accessed on 17 August 2025, https://www.darwinproject.ac.uk/letter/?docId=letters/DCP-LETT-343.xml.
8 Thomas S. Kuhn, *The Structure of Scientific Revolutions* (Chicago: University of Chicago Press, 1962).
9 Darwin, *Autobiography*, 77–78.
10 Darwin, *Autobiography*, 78.
11 Darwin, *Autobiography*, 78.
12 Darwin, *Autobiography*, 78–79.
13 Darwin, *Autobiography*, 78.
14 Darwin, *Autobiography*, 80.
15 Darwin, *Autobiography*, 80.
16 Darwin, *Autobiography*, 80–81.
17 Darwin, *Autobiography*, 81.
18 Darwin, *Autobiography*, 81–82.
19 Darwin, *Autobiography*, 83.
20 Darwin, *Autobiography*, 83.
21 Darwin, *Autobiography*, 83.
22 Darwin, *Autobiography*, 83.
23 Darwin, *Autobiography*, 85–86.
24 Darwin, *Autobiography*, 86.
25 Darwin, *Autobiography*, 86.
26 Darwin, *Autobiography*, 87.
27 Darwin, *Autobiography*, 87.
28 Darwin, *Autobiography*, 87.
29 Darwin, *Autobiography*, 95.
30 Darwin, *Autobiography*, 97.
31 Darwin, *Autobiography*, 100.
32 Darwin, *Autobiography*, 101.
33 Darwin, *Autobiography*, 102.
34 Darwin, *Autobiography*, 104.
35 Darwin, *Autobiography*, 104–107.
36 Darwin, *Autobiography*, 110.
37 Darwin, *Autobiography*, 113.
38 Darwin, *Autobiography*, 113.
39 Darwin, *Autobiography*, 114.
40 Darwin, *Autobiography*, 114–115.
41 Darwin, *Autobiography*, 115–116.
42 Darwin, *Autobiography*, 118.
43 Darwin, *Autobiography*, 119.
44 Darwin, *Autobiography*, 119.
45 Darwin, *Autobiography*, 121.
46 Darwin, *Autobiography*, 121–122.
47 Darwin, *Autobiography*, 122.
48 Darwin, *Autobiography*, 122–123.
49 Darwin, *Autobiography*, 123.
50 Darwin, *Autobiography*, 124.
51 Darwin, *Autobiography*, 124.

52 Fritz Müller, "Darwin Correspondence Project," University of Cambridge, 2022, https://www.darwinproject.ac.uk/fritz-m-ller.
53 Darwin, *Autobiography*, 125.
54 Darwin, *Autobiography*, 126.
55 Darwin, *Autobiography*, 126.
56 Darwin, *Autobiography*, 127–128.

3 Charles Darwin

The Later Decades and Representing Erasmus Darwin

In this autobiography, Darwin continues in this period of a move to the country owing to ill health to speak of publications to the point that the autobiography seems to be an annotated bibliography with characterization and a human touch. Here is an example of a paragraph about a publication that gives a sense of Darwin's technique in this part of his autobiography:

> In the autumn of 1864 I finished a long paper on *Climbing Plants*, and sent it to the Linnean Society. The writing of this paper cost me four months: but I was so unwell when I received the proof-sheets that I was forced to leave them very badly and often obscurely expressed. The paper was little noticed, but when in 1875 it was corrected and published as a separate book it sold well. I was led to take up this subject by reading a short paper by Asa Gray, published in 1858, on the movements of the tendrils of a Cucurbitacean plant. He sent me seeds, and on raising some plants I was so much fascinated and perplexed by the revolving movements of the tendrils and stems, which movements are really very simple, though appearing at first very complex, that I procured various other kinds of Climbing Plants, and studied the whole subject. I was all the more attracted to it, from not being at all satisfied with the explanation which Henslow gave us in his Lectures, about Twining plants, namely, that they had a natural tendency to grow up in a spire. This explanation proved quite erroneous. Some of the adaptations displayed by climbing plants are as beautiful as those by Orchids for ensuring cross-fertilisation.[1]

Asa Gray and Henslow play a role in the genesis of his studies and this paper. Darwin has long been a part of the scientific community at Cambridge but also internationally, Gray being on the faculty of Harvard. Darwin read Gray's work and Gray sent him seeds and that happens to prompt Darwin's attempt to correct a mistake Henslow made at Cambridge.

DOI: 10.4324/9781003651871-4

Another well-known publication also shows Darwin's strategy in this autobiographical narrative:

> My *Descent of Man* was published in Feb. 1871. As soon as I had become, in the year 1837 or 1838, convinced that species were mutable productions, I could not avoid the belief that man must come under the same law.[2]

Man is just another mutable species. Darwin often goes back to the time of his voyage on the *Beagle* to the origin of his ideas, later made into published books. There are narrative and temporal shifts in this autobiographical account. He speaks about his publication in 1876 and the eleven years of experiment that went into it. Soon Darwin adds:

> I hope also to republish a revised edition of my book on Orchids, and hereafter my papers on dimorphic and trimorphic plants, together with some additional observations on allied points which I never have had time to arrange. My strength will then probably be exhausted, and I shall be ready to exclaim 'Nunc dimittis.'[3]

Even in his later years, Darwin is working hard and patiently. After he suffers exhaustion, Darwin will use the phrase, and here he echoes the canticle from the prayer of Simeon in Luke 2:29–32: "now lettest thou depart." He adapts this from his Christian upbringing. One of these projects relates to Darwin's grandfather:

> In 1879, I had a translation of Dr. Ernst Krause's *Life of Erasmus Darwin* published, and I added a sketch of his character and habits from materials in my possession. Many persons have been much interested by this little life, and I am surprised that only 800 or 900 copies were sold. Owing to my having accidentally omitted to mention that Dr. Krause had enlarged and corrected his article in German before it was translated, Mr Samuel Butler abused me with almost insane virulence.[4]

Darwin says he apologized privately to Butler for the omission. Butler seems to have kept on in the controversy and, as noted before, Darwin always tried to avoid controversy. Darwin also says he has sent – on 1 May 1881 – a small book to the printer: "It is the completion of a short paper read before the Geological Society more than forty years ago, and has revived old geological thoughts."[5] Darwin takes up unfinished business, something he considered decades before.

Moving from the publications to a conclusion of his autobiography, Darwin observes:

> I have now mentioned all the books which I have published, and these have been the milestones in my life, so that little remains to be said. I am not conscious of any change in my mind during the last thirty years, excepting in one point presently to be mentioned; nor indeed could any change have been expected unless one of general deterioration.[6]

Darwin was writing about the development of his mind, and he meditates on his publications then on his mind. He says his father died with a clear mind and hopes the same will happen for him. He considers his own state: "I think that I have become a little more skilful in guessing right explanations and in devising experimental tests; but this may probably be the result of mere practice, and of a larger store of knowledge."[7] Darwin sees his continuing skills as being a matter of practice.

The subject of writing is something Darwin considers here, which makes sense given how much Darwin wrote and published:

> I have as much difficulty as ever in expressing myself clearly and concisely; and this difficulty has caused me a very great loss of time; but it has had the compensating advantage of forcing me to think long and intently about every sentence, and thus I have been often led to see errors in reasoning and in my own observations or those of others.
>
> There seems to be a sort of fatality in my mind leading me to put at first my statement and proposition in a wrong or awkward form. Formerly I used to think about my sentences before writing them down; but for several years I have found that it saves time to scribble in a vile hand whole pages as quickly as I possibly can, contracting half the words; and then correct deliberately. Sentences thus scribbled down are often better ones than I could have written deliberately.[8]

The deliberation over a sentence not only helped Darwin to think but it also slowed him down in writing. This way he could see his own errors and those of others. But Darwin qualifies this method by saying that scribbling then deliberating was better than writing deliberately. This way of writing worked for him. He goes into great detail about his method of arranging the material in his book, which, as he mentioned before, is reminiscent of Buckle's way of working. He starts with an outline and uses amplification, portfolio, memorandum, index of other books and abstract – his description showing the complex operation whose end is to retrieve all the information he has collected in his life on the subject on which he is writing.[9]

Another shift is how Darwin's mind has changed regarding the arts:

> I have said that in one respect my mind has changed during the last twenty or thirty years. Up to the age of thirty, or beyond it, poetry of many kinds, such as the works of Milton, Gray, Byron, Wordsworth, Coleridge, and Shelley, gave me great pleasure, and even as a schoolboy I took intense delight in Shakespeare, especially in the historical plays. I have also said that formerly pictures gave me considerable, and music very great delight. But now for many years I cannot endure to read a line of poetry: I have tried lately to read Shakespeare, and found it so intolerably dull that it nauseated me.[10]

As a schoolboy, Darwin delighted in Shakespeare, but, in his later years, Darwin found him dull and nauseating. The poets of his youth no longer gave him pleasure. Darwin cannot even read a line of poetry. Perhaps the educated gentleman who was religious became the scientist who cared less about poetry and culture. Darwin was still reading poetry on the *Beagle* and in the few years after if Darwin remembers correctly. He thinks about art and music: "I have also almost lost any taste for pictures or music.—Music generally sets me thinking too energetically on what I have been at work on, instead of giving me pleasure."[11] One of the themes in the descriptions of the last decades of Darwin's life is ill health or loss of pleasure. Darwin loses the taste for the arts and perhaps even, from time to time, for landscapes themselves: "I retain some taste for fine scenery, but it does not cause me the exquisite delight which it formerly did."[12] The loss of delight recurs in time. Yet, fiction provides some reprieve:

> On the other hand, novels which are works of the imagination, though not of a very high order, have been for years a wonderful relief and pleasure to me, and I often bless all novelists. A surprising number have been read aloud to me, and I like all if moderately good, and if they do not end unhappily—against which a law ought to be passed. A novel, according to my taste, does not come into the first class unless it contains some person whom one can thoroughly love, and if it be a pretty woman all the better.[13]

Darwin blesses novelists even if they are not high order works of imagination because they have given him pleasure and relief. The novels are often read to him. He does not like unhappy endings and he thinks a novel has to have a character or person, preferably "a pretty woman," one can love to be first class. Here is Darwin as a literary, art and cultural critic with his own distinct and changing tastes over the course of his life.

As Darwin is writing largely, in this autobiography, about the development of his mind, he is trying to understand the changes to his mind over the course of his life:

> This curious and lamentable loss of the higher aesthetic tastes is all the odder, as books on history, biographies and travels (independently of any scientific facts which they may contain), and essays on all sorts of subjects interest me as much as ever they did. My mind seems to have become a kind of machine for grinding general laws out of large collections of facts, but why this should have caused the atrophy of that part of the brain alone, on which the higher tastes depend, I cannot conceive.[14]

He finds it odd at his loss of aesthetic taste, which he calls "curious and lamentable." Why, he wonders, can he lose an interest in poetry and still read fiction beyond his interest in science. He sees his mind as becoming a machine that grinds laws out of collected facts, but he cannot understand why this change to his brain atrophied the aesthetic part of his brain. Darwin, the writer, seems to have an aesthetic sense. Clarity is not just clear thinking, but it is also a sense of beauty. Darwin often likes comparison and general observations and not simply seeing himself in isolation:

> A man with a mind more highly organised or better constituted than mine, would not I suppose have thus suffered; and if I had to live my life again I would have made a rule to read some poetry and listen to some music at least once every week; for perhaps the parts of my brain now atrophied could thus have been kept active through use. The loss of these tastes is a loss of happiness, and may possibly be injurious to the intellect, and more probably to the moral character, by enfeebling the emotional part of our nature.[15]

He speaks of an imagined person with a more highly organized mind and supposes that this person would not have suffered this loss of an appreciation of the aesthetic. Darwin imagines living his life again, a life in which he would read poetry and listen to music once a week so that he could keep active the part of his brain that atrophied. He associates a loss of aesthetic tastes with a loss of happiness, a possible injury to the intellect and a probable one to "moral character" by weakening emotions. Darwin takes the aesthetic sense seriously and stresses its importance in terms of well-being, intellect, ethics and feelings, so the question arises whether Darwin thinks that this has affected his scientific mind or his effectiveness as a scientist, as a person studying science. He is obviously good at collecting facts and formulating general laws. He is also interested in possibilities, alternatives and supposition.

This latter part of his autobiography can contain regrets and is also Darwin trying to understand his mind and life, his contribution to science and the possibility that his contribution will endure. The part of the autobiography after his move to the country he centred most, as he said he would, on his publications. He notes that his books, translated into numerous tongues and having had editions in other countries, have sold in England. Some say that success abroad shows "enduring value," but Darwin doubts this but says if so, his "name ought to last for a few years" and he adds: "Therefore it may be worth while for me to try to analyse the mental qualities and the conditions on which my success has depended; though I am aware that no man can do this correctly."[16] He would analyse his own mind, although he sees that to do so correctly is impossible. This statement seems to throw the whole of the autobiography, the discussion of the development of the author's mind, into question.

None the less, Darwin attempts to characterize his mind based on his own observations and answering his own critics. He begins with his own assessment of his mental powers:

> I have no great quickness of apprehension or wit which is so remarkable in some clever men, for instance Huxley. I am therefore a poor critic: a paper or book, when first read, generally excites my admiration, and it is only after considerable reflection that I perceive the weak points. My power to follow a long and purely abstract train of thought is very limited; I should, moreover, never have succeeded with metaphysics or mathematics. My memory is extensive, yet hazy: it suffices to make me cautious by vaguely telling me that I have observed or read something opposed to the conclusion which I am drawing, or on the other hand in favour of it; and after a time I can generally recollect where to search for my authority. So poor in one sense is my memory, that I have never been able to remember for more than a few days a single date or a line of poetry.[17]

Rather than boast about his mind, Darwin sets out its limitations. He is not clever like Huxley, is a poor critic, has a limited ability to follow abstract thought and has a poor memory. His lack of memory he illustrates by how soon he forgets a date or a verse. Darwin almost makes himself an anti-hero, someone who has succeeded against the odds. But, as is Darwin's wont, he balances this frank and almost ironic or satirical self-portrait by turning to his critics:

> Some of my critics have said, "Oh, he is a good observer, but has no power of reasoning." I do not think that this can be true, for the

> *Origin of Species* is one long argument from the beginning to the end, and it has convinced not a few able men. No one could have written it without having some power of reasoning. I have a fair share of invention and of common sense or judgment, such as every fairly successful lawyer or doctor must have, but not I believe, in any higher degree.[18]

Darwin presents the other side, using his critics to address his powers of observation and reasoning, providing *Origin* as an example of extended argument and of reasoning. He also says he had the judgement or common sense of a doctor or lawyer but not more. Thus, Darwin defends himself but limits his claims for his mental capacities or characteristics so as not to make himself at the extreme of extraordinary.

Amplification and illustration are important to Darwin's narrative and self-reflection and self-representation:

> On the favourable side of the balance, I think that I am superior to the common run of men in noticing things which easily escape attention, and in observing them carefully. My industry has been nearly as great as it could have been in the observation and collection of facts.[19]

Darwin sees his observation, collecting facts and industry as uncommon. Darwin highlights his emphasis: "What is far more important, my love of natural science has been steady and ardent. This pure love has, however, been much aided by the ambition to be esteemed by my fellow naturalists."[20] Love of natural science and the esteem of naturalists were the core of Darwin's life as a scientist. He looks back:

> From my early youth I have had the strongest desire to understand or explain whatever I observed,—that is, to group all facts under some general laws. These causes combined have given me the patience to reflect or ponder for any number of years over any unexplained problem.[21]

Understanding, explanation, patience and reflection, Darwin characterizes his own way of looking at and solving problems. Darwin elaborates:

> As far as I can judge, I am not apt to follow blindly the lead of other men. I have steadily endeavoured to keep my mind free, so as to give up any hypothesis, however much beloved (and I cannot resist forming one on every subject), as soon as facts are shown to be opposed to it.[22]

Darwin portrays himself as having an independent mind and following the facts to establish what he thinks is the case. He gives up hypotheses no matter how dear they are to him. Darwin provides an example to illustrate this free mind:

> Indeed I have had no choice but to act in this manner, for with the exception of the Coral Reefs, I cannot remember a single first-formed hypothesis which had not after a time to be given up or greatly modified. This has naturally led me to distrust greatly deductive reasoning in the mixed sciences.[23]

Darwin names the one instance where he was not open to changing a hypothesis, which he usually gave up or modified. He says that he distrusted deduction. The next sentence is not entirely clear: "On the other hand, I am not very sceptical,—a frame of mind which I believe to be injurious to the progress of science; a good deal of scepticism in a scientific man is advisable to avoid much loss of time; for I have met with not a few men, who I feel sure have often thus been deterred from experiment or observations, which would have proved directly or indirectly serviceable."[24] Darwin says he is not "very sceptical," which he thinks is not good for science, but then adds that scepticism is good for scientists. This may be only a matter of degree or an apparent contradiction because he then seems to imply that scepticism can defer scientists from experimenting and observing when it would have been good to do so. Darwin gives an illustration, which he begins thus: "In illustration, I will give the oddest case which I have known. A gentleman (who, as I afterwards heard, was a good local botanist) wrote to me from the Eastern counties that the seeds or beans of the common field-bean had this year everywhere grown on the wrong side of the pod."[25] Darwin then gives at some length the full account of this illustration and provides another instance: "A second case was an account of several varieties raised by the author from several species of Primula, which had spontaneously yielded a full complement of seed, although the parent plants had been carefully protected from the access of insects."[26] The third case involved a hoax.

There are other circumstances that have favoured Darwin. He summarizes what caused his success:

> My habits are methodical, and this has been of not a little use for my particular line of work. Lastly, I have had ample leisure from not having to earn my own bread. Even ill-health, though it has annihilated several years of my life, has saved me from the distractions of society and amusement.[27]

Methodical habits, leisure and even ill health have allowed Darwin to focus on science. He summarizes: "Therefore, my success as a man of science, whatever this may have amounted to, has been determined, as far as I can judge, by complex and diversified mental qualities and conditions."[28] Darwin compresses what he has presented in the autobiography and uses repetition for emphasis:

> Of these the most important have been—the love of science—unbounded patience in long reflecting over any subject—industry in observing and collecting facts—and a fair share of invention as well as of common-sense. With such moderate abilities as I possess, it is truly surprising that thus I should have influenced to a considerable extent the beliefs of scientific men on some important points.[29]

Darwin uses the word "important" twice to stress the significance of his mind and scientific influence. Once more, and here at the end of the autobiography, Darwin reiterates that his own success (which he moderates in the language of modesty) owes itself to "love of science," "unbounded patience," "industry in observing and collecting facts," invention, common sense. He stresses his "modest abilities" and expresses his surprise that he has influenced scientists considerably and "on important points." Darwin represents himself as having modest talents but having achieved great success. After spending many pages amplifying this observation, he telescopes his self-assessment, despite having said that no person can access himself thus. Darwin signed the end on 3 August 1876 having started on about 28 May 1876 and writing his "sketch of my life" about an hour most afternoons.[30] Although this is a disciplined work schedule, Darwin makes it seem like something just done for his family. In some ways, Darwin was well known and his discussion of science and publication would also be of interest to fellow scientists and those interested in science and his work.

I shall end with a brief analysis of Charles Darwin's biographical note on Erasmus Darwin, his grandfather.[31] As Darwin reports, Ernst Krause wrote a sketch of Erasmus Darwin in German published in 1879 in *Kosmos* and gave Darwin and his brother, Erasmus, permission to have the article translated into English.[32] In Darwin's "Preliminary Notice," his views of his grandfather are my focus. Charles says about Erasmus:

> Judging from his published works, letters, and all that I have been able to gather about him, the vividness of his imagination seems to have been one of his pre-eminent characteristics. This led to his great originality of thought, his prophetic spirit both in science and in the mechanical arts, and to his overpowering tendency to theorise and generalise.[33]

Darwin stresses Erasmus' imagination as leading to original thought and a "prophetic spirit" in science and mechanical arts as well as in theorization and generalization. These are concerns for Darwin concerning himself in his own autobiography.

Darwin develops his view of Erasmus Darwin. "Nevertheless, his remarks, hereafter to be given, on the value of experiments and the use of hypotheses show that he had the true spirit of a philosopher. That he possessed uncommon powers of observation must be admitted."[34] Darwin sees in his grandfather two of the qualities he saw in himself in his autobiography – understanding of the value of experiments and hypotheses and talent for observation. Darwin considers his grandfather's mind as he considered his own in the autobiographical sketch: "The diversity of the subjects to which he attended is surprising. But of all his characteristics, the incessant activity or energy of his mind was, perhaps, the most remarkable."[35] Erasmus Darwin has an active mind.

One key motivation for Darwin to write this account of his grandfather is a biography that his family did not think highly of the book:

> In 1804 Miss Seward published her 'Life of Dr. Darwin.' It was unfortunate for his fame that she undertook this task, for she knew nothing about science or medicine, and the pretentiousness of her style is extremely disagreeable, not to say nauseous, to many persons; though others like the book much. It abounds with inaccuracies, as both my father and other members of the family asserted at the time of its publication.[36]

Darwin not only tries to be balanced, but he also notes that Anna Seward, a poet and friend of Erasmus Darwin, does not know about medicine and science and that her style is pretentious, disagreeable and nauseous (he piles up the adjectives against her). Darwin and his family also noted inaccuracies. Darwin listened to his family. Writing to his son George Howard Darwin, Darwin shows that he heeded his own daughter concerning this biographical sketch:

> Here is a bad job; Henrietta thinks my notice of Dr. D very dull,— almost too dull to publish, & I believe that she is right.— I shall be anxious to hear what Eras. thinks. I suppose I must cut it down largely. No one will ever catch me again trying to go beyond my tether.[37]

Writing and design and weighing the evidence are keys to Darwin, who is also connected with and aware of his family's past and present.

My stress here is on Darwin writing about his grandfather's contributions to poetry and science and not really the controversies with Seward or

Samuel Butler or its relation to Krause's contribution on Erasmus Darwin. Darwin tries to be balanced. He shows the praise for Erasmus Darwin's poetry and then chronicles the changes in taste:

> In 1806 an octavo edition of all his poetical works was published in three volumes. I have myself met with old men who spoke with a degree of enthusiasm about his poetry, quite incomprehensible at the present day. Horace Walpole, in his letters repeatedly alludes with admiration to Dr. Darwin's poetry, and in a letter to Mr. Barrett (May 14th, 1792) writes:—'The "Triumph of Flora," beginning at the fifty-ninth line, is most beautifully and enchantingly imagined; and the twelve verses that by miracle describe and comprehend the creation of the universe out of chaos, are in my opinion the most sublime passages in any author, or in any of the few languages with which I am acquainted.'[38]

Although Darwin shows praise for Erasmus Darwin's poetry, he shows how times change in poetry and poetics. Besides Horace Walpole, the writer and son of Robert Walpole, others esteemed Erasmus Darwin:

> Several poets addressed him in complimentary odes, as may be seen in the edition of 1806. Cowper, who, one would have thought, differed in taste from him as much as one man could from another, yet, in conjunction with Hayley, wrote a poem in his honour, beginning:
>
> No envy mingles with our praise,
> Tho' could our hearts repine
> At any poet's happier lays,
> They would, they must, at thine.
>
> Notwithstanding the former high estimation of his poetry by men of all kinds in England, no one of the present generation reads, as it appears, a single line of it. So complete a reversal of judgment within a few years is a remarkable phenomenon. His verses were, however, quizzed by some persons not long after their publication. In the 'Pursuits of Literature,' they are called:
>
> > Filmy, gauzy, gossamery lines.
> >
> > * * * * * *
> >
> > Sweet tetrandrian, monogynian strains.
>
> But the sudden downfall of his fame as a poet was in great part caused by the publication of the well-known parody the 'Loves of the Triangles.'

> No doubt public taste was at this time changing, and becoming more simple and natural. It was generally acknowledged, under the guidance of Wordsworth and Coleridge, that poetry was chiefly concerned with the feelings and deeper workings of the mind; whereas, Darwin maintained that poetry ought chiefly to confine itself to the word-painting of visible objects. He remarks ('Loves of the Plants': Interlude between Cantos I. and II.) that poetry should consist of words which express ideas originally received by the organ of sight. "... And as our ideas derived from visible objects are more distinct than those derived from the objects of our other senses, the words expressive of these ideas belonging to vision make up the principal part of poetic language. That is, the poet writes principally for the eye; the prose writer uses more abstracted terms." It must be admitted that he was a great master of language. In one of the earliest and best criticisms on his poetry it is said no man "had a more imperial command of words, or could elucidate with such accuracy and elegance the most complex and intricate machinery." Byron called him "a mighty master of unmeaning rhyme."[39]

I quoted this passage at length to show its various shifts, and I left out Darwin's footnotes, so this did not become too cumbersome. Darwin wonders why William Cowper and William Hayley, so different in tastes from Erasmus Darwin, wrote a joint ode in praise of Erasmus' poetry. Charles Darwin then reports that no one reads his grandfather's poetry anymore and attributes that in part to a parody of it. Darwin also thinks that a change in taste or aesthetics to a poetry is simpler and more natural – led by Wordsworth and Coleridge – poets Darwin once liked to read as he noted in his autobiography. He also discusses his grandfather's view that words in poetry express the eye – his vision. Moreover, in this ambivalent context, Darwin praises Erasmus for his mastery of words. I am not sure what Byron means by "unmeaning" – is it a compliment?

Darwin's frankness and his weighing pros and cons extend to his assessment of Erasmus Darwin's work in science:

> His first scientific publication was a paper in the 'Philosophical Transactions' for 1757, in which he confutes the view of Mr. Eeles, that vapour ascends through 'every particle, being endued with a, portion of electric fire.' The paper is of no value, but is curious as showing in what a rudimentary condition some branches of science then were.[40]

This first scientific paper makes no contribution to science but has historical interest. Darwin discusses his grandfather in relation to various fields and limits his contribution, often reporting the words of Erasmus and others. Some of the concerns are theories, speculation and experiments to test the theories and Darwin gives Erasmus some credit as some branches of science were not as advanced in the time of his grandfather. Darwin says: "There is a good deal of psychology in the 'Zoonomia,' but I fear that his speculations on this subject cannot be held to have much value."[41] Here, the speculation does not impress Darwin. Erasmus was also a doctor:

> The 'Zoonomia 'is largely devoted to medicine, and my father thought that it had much influenced medical practice in England; he was of course a partial, yet naturally a more observant judge than others on this point. The book when published was extensively read by the medical men of the day, and the author was highly esteemed by them as a practitioner.[42]

The practice of medicine and a book discussing this practice is one of the contributions of Erasmus Darwin. Darwin calls remarkable the statement of his grandfather as something prophetic: "I hope that microscopic researches may again excite the attention of philosophers, as unforeseen advantages may probably be derived from them like the discovery of a new world."[43] There are other instances from the works of Erasmus Darwin. Darwin himself is a textual critic and a close reader. For instance, he notes:

> The 'Phytologia' was published in 1800. It begins with a discussion on the nature of leaf-buds and flower-buds; and the view, now universally adopted, that a plant consists of "a system of individuals" and not merely of a multiplication of similar organs, originated with Darwin, as I infer from Johannes Müller's 'Elements of Physiology.'
>
> Considering how recently the manner in which plants modify and absorb the nutriment stored up in their roots, tubers, cotyledons, &c. has been understood, the following sentence ('Phytologia,' p. 77) deserves notice: 'The digestive powers of the young vegetable, with the chemical agents of heat and moisture, convert the starch or mucilage of the root or seed into sugar for its own nourishment;... and thus it appears probable that sugar is the principal nourishment of both animal and vegetable beings.'[44]

Thus, Darwin can be critical and appreciative of his grandfather. Here is the same analytical and scientific attitude that he uses in the autobiography regarding himself. Darwin tries to understand his own contributions and the limits of them, and he does the same with Erasmus Darwin. There is a sense of balance and fairness in Darwin's autobiographical and biographical work.

Darwin sees Erasmus Darwin as forward looking. As with himself, with his grandfather, he often sees contributions in terms of publications:

> In the 'Phytologia' he shows himself still more clearly a great sanitary reformer. He insists that the sewage from towns, which is now left buried or carried into the rivers, should be removed for the purpose of agriculture; 'and thus the purity and healthiness of the towns may contribute to the thriftiness and wealth of the surrounding country.' 'There should be no burial places in churches or in churchyards, where the monuments of departed sinners shoulder God's altar, ... but proper burial grounds should be consecrated out of towns.' Nearly a century has elapsed since this good advice was given, and it has as yet only partially been carried out.[45]

This is an instance of reform in sanitation and is among Erasmus Darwin's scientific ideas ahead of his time. He also had interests in related areas and some outside his publications. Darwin gives many examples of his grandfather's interest in mechanical inventions:

> One of the subjects which interested Dr. Darwin most throughout his whole life, and which appears little in his published works, was mechanical invention. This is shown in his letters to Josiah Wedgwood, Edgeworth, and others, and in a huge common-place book full of sketches and suggestions about machines. He seems, however, rarely to have completed anything, with the exception of a horizontal windmill for grinding flints, which he designed for Wedgwood, and which answered its purpose. There are schemes and sketches for an improved lamp, like our present moderators; candlesticks with telescope stands so as to be raised at pleasure to any required height; a manifold writer; a knitting loom for stockings; a weighing machine; a surveying machine; a flying bird, with an ingenious escapement for the movement of the wings, and he suggests gunpowder or compressed air as the motive power.[46]

Only one of Erasmus Darwin's inventions was finished and made. Darwin continues to describe other inventions, which show the versatility and imagination of his grandfather. For instance, Darwin says of him: "A speaking machine was a favourite idea, and for this end he invented a phonetic alphabet."[47] Darwin does not lionize his grandfather. There is no boasting, no secular hagiography. He tries to get things right concerning Erasmus Darwin as he had for himself. He stands up for himself, his grandfather and his family but he does not brush negative evidence or criticism aside.

Some of my central concerns in this chapter and in the next chapter are the self-representation of autobiography, the representation of biography and the role of the self and of the family in connection with others. Science has a method but is also part of the familial, social and cultural contexts. These are also similar dimensions in the lives of Asa Gray and Jane Loring Gray and their lives at Harvard University, something I explore in the next chapter.

Notes

1 Charles Darwin, *The Autobiography of Charles Darwin 1809–1882 with Original Omissions Restored*, edited with Appendix and Notes by his grand-daughter Nora Barlow (London: Collins, 1958), 129.
2 Darwin, *Autobiography*, 130.
3 Darwin, *Autobiography*, 133.
4 Darwin, *Autobiography*, 134.
5 Darwin, *Autobiography*, 136.
6 Darwin, *Autobiography*, 136.
7 Darwin, *Autobiography*, 136.
8 Darwin, *Autobiography*, 136–137.
9 Darwin, *Autobiography*, 137–138.
10 Darwin, *Autobiography*, 138.
11 Darwin, *Autobiography*, 138.
12 Darwin, *Autobiography*, 138.
13 Darwin, *Autobiography*, 138–139.
14 Darwin, *Autobiography*, 139.
15 Darwin, *Autobiography*, 139.
16 Darwin, *Autobiography*, 139–140.
17 Darwin, *Autobiography*, 140.
18 Darwin, *Autobiography*, 140.
19 Darwin, *Autobiography*, 140–141.
20 Darwin, *Autobiography*, 141.
21 Darwin, *Autobiography*, 141.
22 Darwin, *Autobiography*, 141.
23 Darwin, *Autobiography*, 141.
24 Darwin, *Autobiography*, 141–142.
25 Darwin, *Autobiography*, 142.
26 Darwin, *Autobiography*, 143.
27 Darwin, *Autobiography*, 144.
28 Darwin, *Autobiography*, 144–145.
29 Darwin, *Autobiography*, 145.
30 Darwin, *Autobiography*, 145.
31 Ernst Krause, *Erasmus Darwin*. Translated from the German by W. S. Dallas. With a Preliminary Notice by Charles Darwin (London: John Murray, 1879).
32 Charles Darwin, "Preface," in *Erasmus Darwin*, iii.
33 Charles Darwin, "Preliminary Notice," in *Erasmus Darwin*, 48.
34 Darwin, "Preliminary Notice," in *Erasmus Darwin*, 49.
35 Darwin, "Preliminary Notice," 49.
36 Darwin, "Preliminary Notice," 70–71.

37 Charles Darwin, Letter to Sir George Howard Darwin, 21 July 1879, Darwin Correspondence Project, "Letter no. 12149," accessed on 21 August 2025, https://www.darwinproject.ac.uk/letter/?docId=letters/DCP-LETT-12149.xml.
38 Darwin, "Preliminary Notice," 92–93.
39 Darwin, "Preliminary Notice," 95–97.
40 Darwin, "Preliminary Notice," 97.
41 Darwin, "Preliminary Notice," 104.
42 Darwin, "Preliminary Notice," 105.
43 Darwin, "Preliminary Notice," 111.
44 Darwin, "Preliminary Notice," 111–112. I do not record Darwin's notes here and elsewhere.
45 Darwin, "Preliminary Notice," 117–118.
46 Darwin, "Preliminary Notice," 118–119.
47 Darwin, "Preliminary Notice," 120.

4 Jane Gray and Asa Gray

Having discussed Charles Darwin and his family in the last section, I shift here to a family who knew the Darwins, that is the Grays. Asa Gray helped to make Charles Darwin known in the United States. Family matters mattered and still do. It is part of the context for scientists and science. The environment is natural and cultural and includes the private and the public and not simply the larger milieu of economics and politics.

Jane Loring Gray (1821–1909), wife of celebrated and renowned Harvard botanist Asa Gray (1810–1888), helped to build up the work and the posthumous reputation of her husband as a leading scientist, an advocate of Charles Darwin and a popular proponent of science in the nineteenth-century United States.[1] In 1859, according to Kuang-Chi Hung, "Asa Gray's paper on the floristic similarities between Japan and the United States was among the earliest applications of Charles Darwin's evolutionary theory in plant geography."[2] Jane Loring Gray left Asa Gray the scientist for others to discuss and wanted to create a portrait of Asa Gray the person. In this chapter, I discuss the Grays in terms of their partnership in science, place that relationship in context and stress the contribution Jane Gray makes to Asa Gray's legacy, including the way she framed her husband's work and reputation after his death. The emphasis here is on the literary, historical, cultural, biographical and autobiographical dimensions of the Grays' work and on the implications that this work has for botany and science then and now. This chapter also stresses the challenges Jane Gray had owing to her gender, her family and social roles and in the face of delicate health. Gray supported her father, her brother and her husband while accomplishing much on her own. She was determined to contribute to shaping Asa's life as well as using fragments of his autobiographical writing as the basis for his biography, and here I seek to shed light on the importance of her life and work and on how the Grays, individually and together, made significant contributions to science, culture and society.

DOI: 10.4324/9781003651871-5

Autobiography, as Linda Anderson observes,

> has… been recognized since the late eighteenth century as a distinct literary genre and, as such, an important testing ground for critical controversies about a range of ideas including authorship, self-hood, representation and the division between fact and fiction.[3]

The shaping of self, shared authorship, the representation of fact and fiction (events and narrative about events) and the mixing of the autobiographical and the biographical are all embodied in the making of the life of Asa in Jane's own writing as well as in the selection and commentary in her edited volumes of Asa's letters. Biography, as Alan Shelston says, "appeals to our curiosity about human personality, and it appeals to our interest in factual knowledge, in finding out 'what exactly happened.'"[4] Jane Gray frames Asa Gray through autobiography, biography and correspondence and is a significant figure in her own right. In 2010, scholars at Harvard University published *Asa Gray at 200* as part of the bicentennial of his birth, and it includes many of the letters quoted in this chapter and some discussion of Jane Gray as well. I focus here on her in relationship to Asa as a way to celebrate the bicentennial of her birth in 2021.[5]

As Janet Browne has observed, the role of correspondence and related documents is important to naturalists, who relied on contact and correspondence to generate a community of experts. As a leading figure in North American botany, Asa Gray found that these letters with botanists and collectors worldwide expanded his influence, and these enduring friendships also improved the collections at Harvard materially. Letters, as Browne also notes, brought Gray and Charles Darwin into contact and friendship, and Gray saw the power of Darwin's *Origin of Species*, advocated for Darwinism in the United States and corresponded with him about natural selection. Whereas Browne sets this correspondence with Darwin in the context of the United States, I am examining the role of Jane Gray in relation to Asa Gray and his correspondence and autobiography as an important part of the contact, letter writing, travel and community at Harvard, nationally and internationally during his life and after his death.[6] My work builds on the work of all those who have discussed Asa Gray and Jane Gray and especially on the contribution of Lisa Ann DeCesare, who focuses on Jane's role in relation to Asa's work and home life and who gives Jane the credit she deserves:

> The Archival Collection of Asa Gray is a rich repository of letters, manuscript material, photographs, notes, articles and accounts concerning Gray's life. This collection, mostly dating from the late 1840s through

the late 1880s[,] was lovingly kept, organized, and cared for by Gray's wife Jane Lathrop Loring Gray.[7]

Jane was a collector and archivist in her own right and her organization helped to arrange and frame Asa's work during his life and after his death, a significant contribution in and of itself. As DeCesare elaborates, "Because of Jane's diligence and foresight we are allowed a rare glimpse of not only Asa Gray the distinguished botanist, the champion of Darwin," but also Asa Gray the "man, husband, and friend."[8] My chapter examines in some detail Jane's relationship to Asa as well as her contribution as an editor and a biographer, particularly in terms of how she frames, selects, orders, interprets and supplements his autobiography and letters, his life in words, his life writing and his writing in life. This chapter also reads Jane Gray's words closely, not simply in relation to those of Asa but in and of themselves.

Before getting to her words in the two posthumous volumes of Asa Gray's correspondence, I provide context mainly from about 1848, when Asa Gray and Jane Loring marry, to 1888, when Asa dies. George Harvey Genzmer's entry on Asa Gray in the *Dictionary of American Biography*, published in 1936, notes Jane Gray's contribution to his work and legacy: "On May 4, 1848, he married Jane Lathrop Loring, daughter of Charles Greeley Loring, who shared sympathetically in his work and survived him to edit his autobiography and letters."[9] Asa Gray was, as Donald H. Pfister says, also instrumental in bringing the study of botany to men and women at Harvard and thereby furthering opportunities for men.[10] Jane's contribution, another example of women making a difference, was acknowledged long ago but needs elaboration and close analysis.

Asa Gray had many talents, one of which was as a writer, and he was part of a tradition, perhaps from Francis Bacon onwards, that used the essay to further natural philosophy, something that Diane Dowdey examines in her analysis of the literary essay in science as a rhetorical form in terms of invention, structure and style.[11] Gray arrived at Harvard with many skills during the summer of 1842 and was impressed with the women in Cambridge. As he wrote to John Torrey, "There's nothing like down East for learned women. Why, even the factory-girls at Lowell edit entirely a magazine, which an excellent judge told me has many better-written articles than the 'North American Review.'" After mentioning how these women help to put their brothers through college, he adds, "Vivent les femmes. There will be no use for men in this region, presently. Even my own occupation may soon be gone; for I am told that Mrs. Ripley is the best botanist of the country round."[12] In addition to his teaching and research duties at

Harvard, Gray became a member of the Boston Society of Natural History, served as corresponding secretary for the American Academy of Arts and Sciences and prepared a series of twelve lectures at the new Lowell Institute for John A. Lowell. In a letter to his father, he wrote,

> I have but just found time to commence the preparation of a course of lectures before the Lowell Institute, which is to commence on the 27th of February, and which will give me plenty of labor and anxiety until they are over.[13]

Gray gave a second series of lectures in 1845 and a third in 1846. It was at one of these Lowell lectures that Asa met Jane, who had been born in Boston to Anna Pierce (Brace) Loring and to Charles Greely Loring, a lawyer, a Harvard College fellow from 1838 to 1857 and a Massachusetts state senator in 1862.[14] Jane also spent time in her mother's hometown of Litchfield, Connecticut. After her mother died in 1836, Jane cared for her father and her two younger siblings, Susan Mary and Charles Greely.[15] As DeCesare notes, few details remain relating to the first meeting of and relationship between Asa Gray and Jane Lathrop Loring.[16] Gray wrote to Jane's father to request his permission to marry her, and Charles Loring responded the following day expressing confidence in Asa and speaking about Jane's view of her duty to her family at home:

> I learned that she is [impressed] with the thought that her duty requires her to remain in her present position—I by no means concur in this opinion, but the subject is one upon which my views would probably have far less influence upon her than any other where my apparent comfort were not so deeply involved.[17]

Jane had a great sense of duty, and her father did not wish her to feel so obliged to care for him that she would neglect herself and her right to marry. That Jane did have this conflict between duty to her family and apprehension about leaving them and making a sacrifice to marry Asa can be seen in a letter Jane wrote to her mother's sister.[18] About two weeks after this first letter, Jane wrote her aunt again to express her own happiness and to state that she let Asa know that her father's happiness and approval of their proposed marriage were a priority for her.[19] Asa wrote to his mother about Jane, her father and their family in positive terms: "She moves in the best, though seldom the most brilliant circles of Boston. Possesses all the usual accomplishments of persons in her station, but is most remarkable for a well-cultivated mind, and for her excellent practical powers."[20] He seems fond of Jane and likes her cultivation and practicality, but they both express some ambivalence about each other, each liking the other's warmth but neither finding the other handsome. Asa appears to distinguish between the best social and intellectual circles and intimates that the Lorings are not part of the latter.

In this later letter to her aunt, Jane records liking Asa's intelligence, enthusiasm, feeling, sympathy and judgement but questioning his excitable nature, adding, "He is quite distinguished as a botanist; & is held in very high esteem by all his friends & acquaintances."[21] Jane looked after George, Asa's younger brother, after he contracted typhoid fever, but on 9 January 1848, while Jane read the Bible to him, George died.[22] On 4 May 1848, Asa and Jane were married: The reception had a choir and was in the Harvard Botanic Garden. The couple lived in the house in the garden. Her sister offered her advice on dealing with the unexpected loneliness that she was feeling. According to DeCesare, Jane did her best to help Asa with his work, but her ill health often held her back, as she expressed to her Aunt Mary:

> I have been sick so much; & now I sometimes lose a day, & that puts one back so much—I really think I sometimes lose time in feeling there are so many things to do; & scarcely knowing where to begin.[23]

Asa and Jane Gray travelled to Europe, where they worked hard, met friends and saw the sights. This was also a way they enriched botany at Harvard and in the United States. They also wrote letters, and, in a letter to her sister, Jane wrote: "Dr. Gray has been kindly aiding me with my journal, taking it up since we were again in England. And I will try before next week to have it all made up to the proper date."[24] In a letter to Elizabeth Lowell, her aunt, dated the following week, she wrote: "I had promised myself this week to bring the journal to present time! but alas this only brings it to Dr. Gray's portion, which you must have plainly seen was his composition not my dictation."[25] This collaboration seems to have involved dictation, even if not in this case, in addition to written composition.

Notable friends they made included Charles Darwin (1809–1882) and his family, whom they visited in England in 1851.[26] In a review of Darwin's *Origin of the Species* (1859), Asa Gray wisely asks:

> Who, upon a single perusal, shall pass judgment upon a work like this, to which twenty of the best years of the life of a most able naturalist have been devoted? And who among those naturalists who hold a position that entitles them to pronounce summarily upon the subject, can be expected to divest himself for the nonce of the influence of received and favorite systems?[27]

Darwin himself came to give Asa the highest praise: "No other person understands me so thoroughly as Asa Gray. If I ever doubt what I mean myself, I think I shall ask him!"[28] Darwin and Asa Gray exchanged ideas, and Jane includes some of their letters in the two posthumous volumes she edited. Jane had her own relationship with the Darwins. Emma Darwin writes Jane Gray on 28 October 1871 and mentions that Charles Darwin is ill. "Charles joins me in kinder remembrances of Professor Gray," adding in a postscript that "Mr Darwin desires me to thank Dr Gray for his kind note, & for your most hospitable invitation to my sons to come to your house, at a time too when Dr Gray was so overwhelmed with business."[29] The scientific relationship between Asa and Charles also involved friendship and a connection between their families.

After Asa's death in 1888, Jane Gray became Asa's biographer and editor. Even prior to her posthumous publication of Asa's letters, Jane wrote two unpublished essays that survive in a notebook housed with her correspondence at Harvard University archives. In them, Jane gives us an idea of the scientist as human, in the sense that she humanizes science and her husband the scientist, beginning her first, untitled essay as follows: "Dr. Gray was particularly alert & active— Of wonderful energy, & endurance." Jane stresses Asa's physical vigour throughout her accounts of their journeys, describing him climbing mountains and going at a pace that was challenging to younger people and adding that "taking his life through he was of unusual vigour & health." Besides describing Asa as impatient, Jane says of Asa's daily intellectual attitude: "A clear idea was what he insisted on." Jane depicts Asa not as a gentle saint but as someone who liked to win an argument no matter what: "He enjoyed a good, strong argument and was often merciless to his opponents— And injurious in turning & twisting what one said against him." Jane represents the intricacy of Asa's character by showing various sides, including this merciless aspect:

> He had great love of wit & fun, & yet he did not enjoy burlesque— His memory for stories was wonderful, & would make a funny turn perhaps unnoticed by his visitors, but if I looked up & saw the little quiver of his nostrils & a gleam in his eyes, I knew there was something that should have been caught.

Jane Gray had a novelistic eye even when regarding her late husband in retrospect.

Jane's portrait of Asa includes a description that gives a fuller sense of the man: "He was of a very hopeful disposition.... And he was extremely fearless." Showing his emotional and psychological aspects, she continues in this vein:

> His feelings were tender & easily touched so that he did not like to hear or read very moving stories or poetry, especially before people— He had the New England dislike of a display of emotion— But the eyes would moisten readily at anything that moved him in any way.

According to this account, Asa felt but did not like to show his feelings. His wife knew by his eyes when he was moved, and he did not like tales and poems that were overtly moving. She continues to discuss his response to music: "He liked simple music, especially singing. And had his favorite hymn tunes." Jane also mentions Asa's interest in other arts:

> But he especially enjoyed fine pictures and Architecture— He carried wonderfully in memory the details of Churches & Cathedrals—recognizing a print or photograph at once of something he had seen perhaps but once thirty or more years before— And was able to make Comparisons— It was his good eye & memory for form.

Asa's eye here, along with his prodigious memory, was for form. He particularly liked paintings and ecclesiastical architecture. She also elaborates on Asa's love of art:

> In pictures he was a good critic, & enjoyed so much his favorite paintings, that I have known him to separate himself from his companions & go through a gallery by himself, because he could not bear to hear the masterpieces he admired, so much criticized—Titian was I think his favorite.

Asa, in this portrait, seemed to want to encounter paintings directly and to form his own interpretation in quiet. His focus also extended from science to art. Jane then tells the story of a young artist in Paris in which she recalls saying to Asa, "Why, what an excellent critic you are." Like a novelist or a biographer, Jane uses dialogue and detail to make the character of her subject come alive in what might be called the drama of the human.

Jane's second essay, entitled "Dr. Gray's Love of Animals," extends his love of plants, nature and art to animals. She begins this section with the story of how, soon after Asa and she had married, their cat had kittens and wandered off for hours, and the kittens nearly starved until Asa fed them milk, after which the kittens mewed and gathered round him in his study. She elaborates on his close connection with these young cats and even personifies them when, as part of their routine, they would give little taps to his shoulder and arm, saying: "It is ten o'clock, we want our supper & to go to bed." Moreover, Jane gives a social sense of Asa and his cats: "His study got many funny names among his friends from the cats." She also says that Asa "had a peculiar skill in handling animals," who seem to have felt comfortable with him. She also remembers their first dog, a Newfoundland puppy: "I always remember the pride with which Dr. Gray ushered him into the gate." She also reports Asa's view of his connection with the dog: "He was most devoted to his master, and Dr. Gray said 'their relationship was more that of brothers, than master and dog.'" Jane talks about Asa at the beach or on the floor playing with this dog and tells more stories about Asa and the cats, including portraits or accounts of each pet and their connection with him, saying that "Dr. Gray was interested in the different characters they showed." Jane also says that Asa would whip and punish the dogs if they misbehaved: For instance, they would dig in the earth in the garden on hot days to cool themselves and needed "much discipline to break them" from this habit. This might have been a time of corporal punishment, not something my pets and I were subject to, so historical difference comes into the interpretation of Jane's biography of Asa and her editing of him. She reveals herself as she reveals him. They used their cultural and historical imaginations as we do, and the future will wonder about us.

The Newfoundland dogs were perhaps like children of the day in a family without children: "They shewed appreciation of fun, did not like to be laughed at, had con- science and good sense[—]Plainly understood the meaning of different words.— Had a choice in the people they liked and were most devoted in affection." In her account, these dogs are sentient and intelligent beings, perhaps personified and humanized through displacement and anthropomorphism, but they are largely loved and respected: Jane loves them as she portrays Asa's love for them. Jane also describes the dogs when Asa came home from church and one of the dogs going to church and yawning there and, satisfied, never again asked to accompany him from Cambridge to the Boston church. Jane has a good sense of humour and irony. To visitors like Asa's younger colleagues or Jane's sister's daughter, the dogs would show affection, and Jane also talks about the skill of the Maltese cat in opening latches. Besides having a talent for such portraits, Jane represents the bond or empathy between Asa and his pets, such as the black and tan terrier Asa did not want as a pet but who endeared himself so much to Asa through warmth and affection and nestled with him in his lap "that Dr. Gray could not resist him, and the dog was perhaps more exclusively devoted to him than any dog we had." This terrier became famous: "He was the dog mentioned by Mr. Darwin, who would wash his face like a cat, licking his paw & then washing his face with it, so that I thought that he might have been suckled by a cat." We might call this Darwin's dog even though it was Gray's dog, and the terrier would call attention to a cat who was lying on a chair it should not have been lying on, would cry when he saw the Grays packing and moved Jane's feelings while Asa dressed the wound on the dog's neck. Jane also tells of Asa's and her return after being away for fifteen months and how the dog was so excited to see Asa again that in a matter of weeks, he was paralyzed and then died. "Poor little fellow, I think the joy killed him," Jane writes, adding that "we truly mourned him." Their next dog was a white and black Japanese terrier without as much of a conscience. They took this dog to a friend's house prior to their trip to Europe in 1880, but he died before they returned.

Their trips, which combined science and leisure, affected their household, human and animal. The last dog the Grays had was a spaniel. Ending this essay, Jane tells about life and death in terms of this dog:

> He was so wretched with when Dr. Gray was taken with his last illness, wandering about the house, & not understanding why his master would not get up & only feebly noticed him, that kind friends took him home and he lives in a home where nothing happier could be asked for a dog; & where now ever he still has a cordial greeting for his old mistress.
> (crossed-out word in original)

The description of the dog and of Asa makes the mortal illness poignant, and, after Asa's death, Jane and the dog are left to greet each other when she visits given that he no longer lives in the Gray household. This story is one of love across species, something Darwin, Asa and Jane seemed to have understood. Science occurs amid friendship and love and in the context of nature, plant and animal.

It is remarkable that Jane considered it important to show this side of an eminent scientist and public figure and to provide a private portrait of her husband not only as a person but also as someone whose work she had helped to further. Asa Gray was also someone who, in addition to his accomplished work as a scientist, wrote about science in a way that appealed to general readers. His memory and legacy were in the hands of Charles Eliot (the president of Harvard), Jane Gray and William Gilson Farlow (his protégé at Harvard) as well as others in the seven years after his death and later in the hands of people such as Hunter Dupree. Jane, as DeCesare notes, helped to "preserve and chronicle" Asa's "life's work":

> She organized his correspondence, manuscripts, and papers and deposited them at Harvard. She transcribed and edited a book of important Gray correspondence titled *Letters of Asa Gray* (1893). She then sent gift copies of the book to major university and garden libraries all over the world.[30]

The public response to the death of Asa Gray is also significant. The meeting of the American Academy of Arts and Sciences on 13 June 1888 "was devoted to a commemoration of the late Asa Gray" and resolved that "a copy of these resolutions be sent to Mrs. Gray, his helpful companion in his life's work, with the respectful sympathies of the Fellows of the Academy."[31] Although this issue records what others say about Asa, the president emphasizes the key role of Eliot in the commemoration. "The life of Asa Gray always seemed to me a singularly happy one," Eliot declares, starting with the person and not the scientist, but the theme of happiness unites the man and the scientist:

> These years fell at a most fortunate period; for the continent was just being thoroughly explored, and its botanical treasures brought to light. Dr. Gray's labors therefore cover the principal period of discovery and of accurate classification in American botany. Merely to have one's intellectual life-work make part of a structure so fair and lasting, is in itself a substantial happiness.

Part of this happy life involved Asa's connection to nature: "His pursuit was one which took him out of doors, and made him intimate with Nature in all her moods."[32] For Eliot, the life of Asa Gray as a botanist had four aspects of joy: "All these are elements of happiness,—love of nature, acquaintance with the wide earth, congenial intercourse with superior minds, and abiding friendships." Being in nature and among friends with good minds was part of the happy life of this botanist. Eliot elaborates: "Although Dr. Gray had no children, his domestic experience was one of rare felicity. His life illustrated a remark of his friend Darwin, that with natural history and the domestic affections a man can be perfectly happy."[33] Eliot appeals to Darwin in seeing happiness between natural history and domestic feelings.

Eliot also stresses Asa's accomplishment: "His reputation was larger than that of a specialist; he was recognized as a clear thinker on philosophical and religious themes, a just and sagacious critic, and a skilful and vigorous writer." Eliot's Gray is a botanist, a thinker, a critic and a writer. Eliot specifies Asa's contribution to science:

> For many years past he could not but know that he had made the largest and most durable contribution to American botanical science which had ever been made, and that he had done more than any other man to diffuse among his countrymen a knowledge of botany and a love for it.

Asa Gray was the pre-eminent botanist in the United States and the last words of Eliot's commemoration or tribute – "heartfelt content" – regarding

Asa's service to botany and to Harvard's department in that field returns to Eliot's consistent theme in this portrait.[34]

In April 1889, Farlow paid tribute to Asa Gray, saying, among other things, that Asa "had made a favourable impression on President Quincy, of Harvard College, and soon after his return to New York he was offered the newly endowed Fisher Professorship of Natural History at Harvard, which he accepted." Farlow tells of Asa moving to Cambridge in 1842 and observes that from the time of Asa's and Jane's marriage "his life ran smoothly on. He was now able to devote himself wholly to botany and the building up of the botanical establishment at Harvard University, which he may be said, in fact, to have created." Farlow recognizes Jane and offers details about Asa's importance to the country: "From Cambridge his influence extended in ever-widening circles until, as time went on, he became the one man to whom American botanists looked for advice, sympathy, and inspiration." After describing Asa's honours, Farlow turns to his death:

> For he was still as fresh, as active, as cheerful as ever when he returned from his last European journey, in October, 1887, and turned with renewed zeal to his work on the *Flora,* which he was destined never to complete, for, on the morning of November 28, he was stricken with paralysis and lingered in a partially conscious condition until the evening of January 30.[35]

Here, Farlow gives the "zeal" of Asa's life as a botanist and the sudden stroke and lingering death.

Farlow sums up Asa's unique contribution:

> The life of Asa Gray fills some of the brightest pages in the history of American science. He occupied a place in the public esteem which no other botanist had ever held, and his death leaves a gap which cannot be filled. It is one of those rare cases where the right man appeared at the right moment.[36]

For Farlow, Asa represents timing and "public esteem" as the pre-eminent American botanist. Asa's "special studies" prepared him to favour Darwin's *Origin of Species*, and

> Darwin's views were already known to Gray, for the two had been in intimate correspondence since 1855, and Gray was one of the three persons to whom advance sheets of the Origin of Species had been sent before its publication. To Gray there was nothing alarming or irreligious in the theory of evolution.[37]

As a religious scientist, Asa defended and advocated for Darwin. Style is important to scientists of the time (Gray once corrected one of Darwin's sentences) and Farlow says, "In his descriptive works Gray's style is marked by clearness and a singularly happy use of short, expressive sentences, and his general and popular essays are written in a style which would do credit to any cultured literary man."[38] For Farlow, Asa was literary and scientific, his writing style clear.

In the prefatory note to the first of the two posthumous volumes of Asa Gray's letters edited by Jane Loring Gray, Jane offers a rationale for this edition: "It has been my aim, in collecting and arranging the 'Letters' from Dr. Gray's large correspondence, to show, as far as possible in his own words, his life and his occupation." Jane is explicit about her goal, which is to arrange the letters in such a way as to allow Asa to tell about his life and work "in his own words." In framing her editorial work in this way, Jane effaces herself to showcase Asa. As she adds, "The greater part of the immense mass of letters he wrote were necessarily purely scientific, uninteresting except to the person addressed; so that many of those published are merely fragments, and very few are given completely." Asa the person is Jane's emphasis, and she does not want to get into the specifics of scientific and professional correspondence but is radical in selecting what she thinks will interest readers of the volumes and not the original reader or addressee. Her aim is to make the books appeal to a broader readership, so she has a rhetorical imperative, taking into consideration her audience, and she elaborates on her emphasis:

> I have made no attempt to estimate his scientific or critical labors, for they are sufficiently before the world in various printed works; but something of the personality of the man and his many interests may be learned from these familiar letters and from even the slight notes.

Jane continues to stress personality and interests: Asa "the man." She shifts from letters to life writing:

> Dr. Gray began an Autobiography, but went no further than to give a brief sketch of his early life. This fragment is placed, with some notes illustrative of the early conditions in which his youth was passed, at the beginning of the work.[39]

Jane begins the volumes with the autobiographical fragment of his early life, giving it pride of place as she helps to shape Asa's life in his own words: the life of a scientist without the science or at least without its technical matter. She ends the brief note with thanks for the help of friends, including Farlow, and the brevity of Jane's note is self-effacing but does the work. About five years and four months after Asa's death, Jane writes about Gray's botany from the Botanic Garden. They lived in the Garden House together, a private and public space in the heart of Harvard and Cambridge.

Jane's notes and selections help to shape Asa as a person, and she combines his words to meld autobiography and biography through letters to depict his life for readers. In his autobiography, Asa portrays himself as a reader and mentions how he first heard Walter Scott read aloud in an inn and how he would borrow novels from a library in Clinton. To this, Jane adds in a note that "in later life the novels were always saved for long journeys," mentioning that he liked Scott and George Eliot:

> Scott always remained the prime favorite, and his last day of reading, when the final illness was stealing so unexpectedly and insidiously on, was spent over *The Monastery*, which he had been planning to read on his homeward voyage in 1887.[40]

The last day of reading before his death was of Scott, so between Asa and Jane, we learn that Asa had been drawn to Scott at his first reading of novels and ended his reading with him. The site of Asa's death is something that Jane, Farlow and others revisit. Asa's autobiography ends in 1843 and is supplemented by Jane's headnote to Chapter 2, "Early Undertakings. 1831–1838":

> Dr. Gray's autobiographical fragment closes abruptly, and is valuable chiefly for the glimpse which it gives of his ancestry and his boyhood. He kept no diary, but he carried on a voluminous correspondence, and his letters thus contain a record of his hard-working, eager life.[41]

Jane interprets the autobiography in terms of family background and childhood and, given that Asa did not keep a diary, she selects letters to help record his early life.

Furthermore, Jane builds on the autobiography with words from the letters that she selects, so that the editor/interpreter becomes the co-writer of Asa's life. She says of the letters: "The earliest tell of the struggle for position, his doubts if his loved science could furnish him a maintenance, and his resolution to make any sacrifice if he could devote himself to its study."[42] Jane derives from these letters that Asa made personal sacrifices in order to study science and says that he forgot his early hardships.

After leaving Fairfield Medical College he divided his years between teaching in Bartlett's school in Utica (some of his old pupils still recall his field excursions with his class, and his eager delight in the search after plants), in journeys botanical and mineralogical, and in some shorter and longer stays in New York, where for a good portion of the time he was a member of Dr. John Torrey's family.[43]

Asa's delight in plants is evident here, and so is his time with his mentor in botany, John Torrey, and his family. Portraits of others such as John Carey, also a distinguished botanist, are likewise part of Jane's account.

In Chapter 3, "First Journey in Europe. 1838–1839," Jane writes: "It has been deemed expedient to give a somewhat fuller narrative of Dr. Gray's first visit to Europe than of his subsequent ones." Jane is justifying her selection: "It was then that he formed many personal acquaintances which ripened into lifelong friendships, and received his first impressions of scenes in nature and art which were to become very familiar." Friendships, art and nature are important for Asa the scientist and Asa the man. Jane explains Asa's method and hers: "His letters home took the form of a very detailed journal, and it is in extracts from this journal, supplemented by letters to other friends, that this narrative consists."[44] Jane's selection from Asa's voluminous journal and letters becomes important as a biographical supplement to his autobiography. In Chapter 4, entitled "A Decade of Work at Home. 1840–1850," Jane writes, "On Dr. Gray's return from Europe, the University of Michigan not yet needing his ser- vices, he settled in New York to work on the 'Flora of North America.'"[45] She also specifies that "in 1841 [Asa] made his first journey to the mountains of North Carolina, of which he wrote an account in the 'American Journal of Science' in the form of a letter to Sir William Hooker" and how he and his friends Dr. Torrey and Dr. Engelmann sent collectors on expeditions, explorations and boundary surveys west of the Mississippi and had "difficulties of communication" owing to restrictive postage regulations.[46]

The second volume opens with Chapter 5, "Second Journey in Europe.— Correspondence. 1850–1859," in which Jane notes that

> Dr. Gray sailed for England with Mrs. Gray in a sailing packet [on] June 11, 1850. The steamers made regular trips, but the fine packets were still running, and it was thought desirable to try the longer voyage for Mrs. Gray's health,

adding that Asa saw old friends and made new ones.[47] In Chapter 6, "Letters to Darwin and Others. 1860–1868," Jane calls attention to her method of selection:

> Dr. Gray's letters to Dr. Darwin previous to 1862 have been destroyed, save the one dated January 23, 1860, which was published in Darwin's

> 'Life and Letters,' and is here reproduced for the convenience of the reader, as well as Dr. Gray's letter of January 5, 1860, to Dr. Joseph D. Hooker, also published in Darwin's 'Life and Letters.'[48]

Concerning the friendship between Darwin and Asa, Jane gives information about the manuscripts and editing while continuing thus:

> The original letters to Darwin later than 1862 have been more or less injured, apparently by the ravages of mice, so that in copying them it has sometimes been necessary to supply missing words. Where these are not obvious, the supposed words are enclosed in brackets.[49]

This description is reminiscent of Philip Sidney's description of historians as "loaden with old mouse-eaten records."[50] Jane adds that the letters include those from the Civil War, into which "Dr. Gray threw himself with all his earnestness" and even, despite his age, "enlisted in Cambridge to guard the State Arsenal there."[51]

Apparently, Darwin and Gray had some differences over the Civil War, but their focus on science seems to have got them through those differences.

In Chapter 7, "Travel in Europe and America. 1868–1880," Jane mentions that Asa "made his fourth journey to Europe in the fall of 1868. He landed in September, and went at once to Kew, where he remained most of the time at work in the herbarium until November."[52] Kew is a key place for plants and for botanists. Jane and he spent time with Mr. Church in Whatley (in Somersetshire), which Jane describes as "one of the loveliest parts of rural England," before moving on to Down, where they spent time with Darwin, Dr. and Mrs. Hooker, Professor Tyndall and others.[53] As Jane wrote, "Those were days never to be forgotten."[54] Speaking of herself in the third person, Jane has not only made herself part of the narrative as a person or character in it but also put in her own feelings and response: "In November, Dr. and Mrs. Gray joined some family friends in Paris, with whom they went to Egypt and passed the winter on the Nile, taking the longest vacation, Dr. Gray said, he had ever enjoyed."[55] They mixed work and pleasure in Paris and Egypt and then "passed through Italy, Switzerland, and Germany, where old botanical acquaintances were renewed, and some persons seen whom he had known only by correspondence. In England he again worked at Kew, and repeated the visits at Whatley and Down."[56] Kew Gardens is a recurrent theme of these European visits. In Chapter 8, "Final Journeys and Work. 1880–1888," Jane writes,

> Dr. Gray sailed for Europe with Mrs. Gray early in September, 1880. He went especially to study herbaria for his new volume of the 'Synoptical

> Flora,' and saw almost every collection of importance, giving especial attention to the subject of asters.[57]

This is a portrait of Asa working hard in his seventies, travelling with Jane. In autumn, he worked in France and Spain, "and in Madrid he looked over the herbarium there. He declared nobody had ever had so many asters pass through his hands as he had had!"[58] Jane uses the third person and indirect speech to provide an image of Asa amid the asters, proceeding to a description of a winter "spent in hard work in the Kew Herbarium" and enjoying "heartily in spring a journey through Italy with his friends Sir Joseph and Lady Hooker, returning to Kew to spend the summer, at work in the herbarium," then sailing "again for home in October, 1881, landing early in November."[59] These trips aided Asa in botany among other botanists and in herbaria in Madrid, Kew and elsewhere.

Included as an appendix to the second volume is a section entitled "Dr. Gray's Will," according to which Asa "left to the herbarium the proceeds of all his copyrights."[60] Even after Asa's death, Jane reinforces Asa's legacy by emphasizing the magnitude of his contributions to botany at Harvard and in the United States:

> His strong belief in the importance of a large and well-kept herbarium, as an establishment indispensable not only to the development of botanical studies at Harvard but also to the diffusion throughout the whole country of a knowledge of its flora, is shown by this bequest, and also by the active efforts he was constantly making in its behalf during his life and by the personal sacrifices to which he cheerfully submitted, that the herbarium might profit.[61]

Jane draws a verbal picture here of Asa as a botanist who has built up botany at Harvard, in the United States and internationally and as someone who made "personal sacrifices" to this end. Jane amplifies this view: "The reputation of Dr. Gray at home and abroad naturally served to draw to the herbarium large numbers of specimens," and she says that he was willing to give Harvard his collection should it construct a building to keep them safe from fire and destruction.[62] She gives a partial list of his publications to end the second of two volumes.

Asa Gray has had a notable reception over the past eighty years or so. After the Second World War, Hunter Dupree, then a graduate student at Harvard, revived and extended our understanding of Asa Gray. A colleague of Thomas Kuhn at Harvard and Berkeley, who was likewise a key historian of science, Dupree published in 1959 a biography of Asa Gray that remains a significant moment in the history of science, in the history of biology and of botany and in studies of Asa Gray.[63] After expressing some reservations, Donald Fleming proclaims in a review of Dupree's biography that "there is no better biography of an American scientist."[64] Dupree has inspired others since his groundbreaking work on Asa Gray, as Judith Warnement stated so well in 2010: "On the occasion of Asa Gray's bicentennial we express our deepest gratitude to Hunter Dupree for keeping Gray's story alive."[65] My chapter, I hope, has extended this knowledge that Eliot, Farlow, Dupree and others considered here have forged about Asa Gray, but to stress the work of Jane Loring Gray, whom Eliot, Farlow and others mentioned positively in passing and to whom Dupree and more recently DeCesare have called attention. It may be commonplace to think of men framing the reputations of women, but Jane Gray was the foremost framer of Asa Gray's legacy soon after his death. Eliot and Farlow, members of Harvard, might have done some of this work along with botanists in the United States, Europe and elsewhere, but Jane Gray was central in remembering Asa Gray in death as he was in life.

Jane Gray framed and edited the man in his domestic and other spaces. She even became co-author of his autobiography and the editor of his correspondence. Jane set out to make Asa human and at home in nature among plants and animals. She wanted to show this great botanist and public advocate and explainer of science in his private and public dimensions, to humanize science and to show other humans that this endeavour and this scientist were worthy of their empathy and attention. Jane Loring Gray did a service to Asa Gray, to botany and science and to us all then and now. The next chapter discusses another family, the family of Roland Thaxter, who worked with Farlow at Harvard.

Notes

1 Lisa Ann DeCesare, "Jane Lathrop Loring Gray (1821–1909) and the Archives of the Gray Herbarium," *Harvard Papers in Botany* 15 (2) (1 December 2010): 221–230. https://doi.org/10.3100/025.015.0205.
2 Kuang-Chi Hung, "Finding Patterns in Nature: Asa Gray's Plant Geography and Collecting Networks (1830s–1860s)," Harvard University, PhD dissertation, 2013, iii; see also 3, 5–6, 400, https://nrs.harvard.edu/urn-3:HUL.InstRepos:11181178.
3 Linda Anderson, *Autobiography* (2nd ed.) (London: Routledge, 2011), 1.
4 Alan Shelston, *Biography* (1977, London: Routledge, 2018), 3.
5 Donald H. Pfister, "Introduction," *Asa Gray at 200*. Harvard U Herbaria and Libraries, 2010. https://huh.harvard.edu/book/asa-gray-200.
6 Janet Browne, "Asa Gray and Charles Darwin: Corresponding Naturalists," *Harvard Papers in Botany*, 15 (2) (2010): 209. https://doi.org/10.3100/025.015.0204; see Anderson Hunter Dupree, "The First Darwinian Debate

in America: Gray Versus Agassiz," *Daedalus: Proceedings of the American Academy of Arts and Sciences* 88 (3) (Summer, 1959): 560–569. https://www.jstor.org/stable/20026523.

7 DeCesare, "Jane Lathrop Loring Gray," 221

8 DeCesare, "Jane Lathrop Loring Gray," 221.

9 George Harvey Genzmer, "Asa Gray," in *Dictionary of American Biography*, edited by Allen Johnson and Dumas Malone. Vol. 7. (New York: Charles Scribner's Sons, 1936), 512.

10 Donald H. Pfister, "Asa Gray and Harvard Summer School," *Harvard Papers in Botany* 15 (2) (2010): 307–308. https://doi.org/10.3100/025.015.0209; see also George E. Gifford, Jr., "Edward Frederick Leitner (1812–1838) Physician-Botanist," *Bulletin of the History of Medicine* 46 (6) (1972): 568. https://www.jstor.org/stable/44450833. See Donald H. Pfister, "Introduction," *Harvard Papers in Botany* 15 (2) (2010): 203–204. https://doi.org/10.3100/025.015.0202.

11 Diane Dowdey, "Literary Science: A Rhetorical Analysis of an Essay Genre and Its Tradition," University of Wisconsin, PhD dissertation, 1984, ii, 34–35, 39.

12 Asa Gray, Letter to John Torrey, 25 July 1842. Asa Gray correspondence, Botany Libraries, Gray Herbarium Library, Harvard University, qtd. in DeCesare 221, 222. For all the original letters, see Asa Gray, Correspondence files of the Gray Herbarium, 1820–1904. Botany Libraries, Gray Herbarium Library, Harvard University. See also Asa Gray, Papers, 1830–1953, Botany Libraries, Gray Herbarium Library, Harvard University.

13 Asa Gray to Mary Gray, 18 November 1843; see also DeCesare 222.

14 DeCesare, "Jane Lathrop Loring Gray," 222.

15 DeCesare, "Jane Lathrop Loring Gray," 223.

16 DeCesare, "Jane Lathrop Loring Gray," 224.

17 Charles G. Loring, Letter to Asa Gray, 4 March 1847. Charles G. Loring Papers, Houghton Library, Harvard University, qtd. in DeCesare 224.

18 Jane Loring, Letter to Mary Pierce, 25 April 1847, Loring Family Correspondence, Lichfield Historical Society; see also DeCesare 225.

19 Jane Loring, Letter to Mary Pierce, 9 May 1847, Loring Family Correspondence, Lichfield Historical Society; see also DeCesare 225.

20 Asa Gray, Letter to Roxana Gray, 6 May 1847, Asa Gray correspondence, Botany Libraries, Gray Herbarium Library, Harvard University, qtd. in DeCesare 221.

21 Jane Loring, Letter to Mary Pierce, 20 January 1849. Loring Family Correspondence, Lichfield Historical Society, qtd. in DeCesare 225

22 See A. Hunter Dupree, *Asa Gray, 1810–1888* (Cambridge, MA: Belknap Press of Harvard University Press, 1959), 181; DeCesare 226.

23 Jane Loring to Pierce, 20 January 1849, qtd. in DeCesare 226–227.

24 Jane Loring Gray, Letter to Susan Loring Jackson, 3 July 1851. Asa and Jane Gray Travel Correspondence, 1838–1869, Gray Herbarium Library, Harvard University. See Gray, Jane Loring. Autograph collection, 1563–1908. Gray Herbarium Library, Harvard University; Jane Loring Gray, Correspondence, 1848–ca. 1895, Gray Herbarium Library, Harvard University.

25 Jane Loring Gray, Letter to Elizabeth Lowell, 11 July 1851. Asa and Jane Gray Travel Correspondence, 1838–1869, Gray Herbarium Library, Harvard University.

26 Dupree, *Asa Gray*; Browne 209–210, 213.

27 Asa Gray, *Darwiniana: Essays and Reviews Pertaining to Darwinism.* 1876, edited by A. Hunter Dupree (Cambridge, MA: Belknap Press of Harvard University Press, 1963), 8.
28 Qtd. in Browne 214.
29 Emma Darwin, Letter to Jane Loring Gray, 28 October 1871, Jane Loring Gray correspondence, Botany Libraries, Gray Herbarium Library, Harvard University.
30 DeCesare, "Jane Lathrop Loring Gray," 228.
31 "Proceedings," *Proceedings of the American Academy of Arts and Sciences*, vol. 24, 1888–1889, 406–407, https://www.jstor.org/stable/20021574.
32 "Proceedings," 408.
33 "Proceedings," 409.
34 "Proceedings," 409.
35 William Gilson. Farlow, "Memoir of Asa Gray, 1810–1888: Read before the National Academy, April 17, 1889," *Biographical Memoirs* 3 (3) (1985): 166. http://www.nasonline.org/publications/biographical-memoirs/memoir-pdfs/gray-asa.pdf.
36 Farlow, "Memoir of Asa Gray," 166.
37 Farlow, "Memoir of Asa Gray," 171.
38 Farlow, "Memoir of Asa Gray," 173.
39 Jane Loring Gray, editor, *Letters of Asa Gray* (Boston, MA: Houghton, Mifflin and Company, 1894), 2 vols., 1: n. pag.).
40 Jane Loring Gray, editor, *Letters of Asa Gray*, 1: 12 n1.
41 Jane Loring Gray, editor, *Letters of Asa Gray*, 1: 29.
42 Jane Loring Gray, editor, *Letters of Asa Gray* 1:29.
43 Jane Loring Gray, editor, *Letters of Asa Gray*, I: 29.
44 Jane Loring Gray, editor, *Letters of Asa Gray*, 1: 85.
45 Jane Loring Gray, editor, *Letters of Asa Gray*, 1: 272.
46 Jane Loring Gray, editor, *Letters of Asa Gray*, 1: 272.
47 Jane Loring Gray, editor, *Letters of Asa Gray*, 2: 369.
48 Jane Loring Gray, editor, *Letters of Asa Gray*, 2: 454.
49 Jane Loring Gray, editor, *Letters of Asa Gray*, 2: 454.
50 Philip Sidney, *The Defense of Poesy Otherwise Known as an Apology for Poetry*, edited by Albert S. Cook (Boston, MA: Ginn and Company, 1901), 13.
51 Jane Loring Gray, editor, *Letters of Asa Gray*, 2: 453.
52 Jane Loring Gray, editor, *Letters of Asa Gray*, 2: 565.
53 J. D. H., "Obituary Notices of Fellows Deceased," *Proceedings of the Royal Society of London* 46 (1889): xv–xviii. https://www.jstor.org/stable/115081.
54 Jane Loring Gray, editor, *Letters of Asa Gray*, 2: 565.
55 Jane Loring Gray, editor, *Letters of Asa Gray*, 2: 565.
56 Jane Loring Gray, editor, *Letters of Asa Gray*, 2: 565.
57 Jane Loring Gray, editor, *Letters of Asa Gray*, 2: 701.
58 Jane Loring Gray, editor, *Letters of Asa Gray*, 2: 701.
59 Jane Loring Gray, editor, *Letters of Asa Gray*, 2: 701.
60 Jane Loring Gray, editor, *Letters of Asa Gray*, 2: 816.
61 Jane Loring Gray, editor, *Letters of Asa Gray*, 2: 816.
62 Jane Loring Gray, editor, *Letters of Asa Gray*, 2: 816.
63 See A. Hunter Dupree and Thomas S. Kuhn, "Teaching the History of Science: University of California at Berkeley," *Isis* 49 (1958): 172–173. https://doi.org/10.1086/348670; Dupree, "First"; Karl Hufbauer, "From Student of Physics to Historian of Science: T.S. Kuhn's Education and Early Career,

1940–1958," *Physics in Perspective* 14 (4) (2012): 421, 438, 440, 443, 449, 452–453, 460. https://doi.org/10.1007/s00016-012-0098-5; Frank N. Egerton, "History of Ecological Sciences, Part 44: Phytopathology during the 1800s," *Bulletin of the Ecological Society of America* 93 (4) (2012): 326–327. https://doi.org/10.1890/0012-9623-93.4.303.

64 Donald Fleming, "Review of *Asa Gray, 1810–1888*," edited by A. Hunter Dupree, *American Historical Review* 65 (4) (1960): 925. https://doi.org/10.2307/1849458.

65 Judith Warnement, "Reflections on a Life with Asa Gray: A Conversation with Hunter Dupree," *Harvard Papers in Botany* 15 (2) (2010): 320. https://doi.org/10.3100/025.015.0211.

5 The Thaxters

Nature, Science and Art

In *An Island Garden* (1894), Celia Thaxter includes the "PLAN OF GARDEN WITH LIST OF FLOWERS," which lists fifty-seven flowers, number 2 being "Asa Gray Nasturtiums."[1] Here is one connection between the Grays and the Thaxters. A love of nature also connects them. Celia is not only a poet and writer but also a conserver and chronicler of nature. In this plan, Celia details her garden and what borders it: "The garden is 50 ft. long by 15 ft. wide, and is surrounded by a border of all sorts of mixed flowers. A bank of flowers at the southwest corner slopes from the garden fence."[2] I shall return to this book below.

I really owe my interest in the Asa Gray, William Farlow, Roland Thaxter and their associates and families to Donald H. Pfister.[3] He and I explored the possibility of an exhibition or display at Houghton Library at Harvard and, by April 2019, we put together this idea for "The Thaxter Family and the World of Nature and Culture: Science, Art and Community at Harvard, in New England and Beyond," but the library had some backlog and then came the pandemic and all changed.[4] In this chapter, I will, among other things, present some of these ideas, in the context of my own discussion of science and culture and the role of families. Thaxter studied with Farlow who studied with Gray at Harvard.

Donald H. Pfister, Asa Gray Professor of Systematic Botany and Curator of the Farlow Library and Herbarium Emeritus at Harvard, thinks that humanists should know more about the Thaxters and he highlights some of their contributions. He also reminded me of Celia's garden, which she describes in *An Island Garden*, perhaps her most popular prose work and that is still in print, illustrated by Childe Hassam, one of the frequent visitors during the summer to the hotel on the Appledore Island. Pfister also stressed that in the context of art and culture, Celia entertained many artists, writers, musicians on Appledore.[5] Of Celia Thaxter, he says:

> she was a poet perhaps first and foremost, but her prose writing is equally important. An Island Garden and her connection with Childe

DOI: 10.4324/9781003651871-6

> Hassam is well known; he illustrated the book. The book explains her garden and her connection to the garden and her other work. She made money painting dinnerware, china, as many Victorian ladies did, but her first hand acquaintance with the plants made a difference. She was an advocate for preserving birds because she realized women's fashion was driving some birds toward extinction.[6]

Donald H. Pfister reminds us of the many dimensions of Celia's work. The Metropolitan Museum in New York has a number of works of Childe Hassam, including "Celia Thaxter's Garden, Isles of Shoals, Maine" (1890), which the museum describes thus:

> This painting is one of the finest of a series of works that Hassam made during summers in the 1890s on Appledore Island, one of the Isles of Shoals, which lie ten miles east of Portsmouth, New Hampshire. This series portrays the sumptuous wildflower garden cultivated by his friend, poet Celia Thaxter, a garden that provided a marvelous contrast to the rugged terrain of the island itself. In this painting, vibrant red poppies entangled in lush green foliage introduce a view of bleached Babb's Rock. The painting shows Hassam at the height of his creativity as an American Impressionist.[7]

Celia's commitment to her garden, to plants and animals and to conservation is important and is something that some of the literary audience might have overlooked, although conservation and ecology and what is now called ecological criticism have been concerns of humanists and literary scholars for some time now. Celia begins *An Island Garden*:

> OF all the wonderful things in the wonderful universe of God, nothing seems to me more surprising than the planting of a seed in the blank earth and result thereof. Take a Poppy seed, for instance: it lies in your palm, the merest atom of matter, hardly visible, a speck, a pin's point in bulk, but within it is imprisoned a spirit of beauty ineffable, which will break its bonds and emerge from the dark ground and blossom in a splendor so dazzling as to baffle all powers of description.[8]

Celia extols the beauty of seeds, an example of her love of nature, and uses the topos of inexpressibility in which she expresses the inexpressible.[9] Not only does Celia describe flowers and nature in her poetry and her writing, but she also devotes her time to observing and growing flowers and conserving nature. Another passage illustrates her devotion:

> He who is born with a silver spoon in his mouth is generally considered a fortunate person, but his good fortune is small compared to that

> of the happy mortal who enters this world with a passion for flowers in his soul. I use the word advisedly, though it seems a weighty one for the subject, for I do not mean a light or shallow affection, or even an æsthetic admiration; no butterfly interest, but a real love which is worthy of the name, which is capable of the dignity of sacrifice, great enough to bear discomfort of body and disappointment of spirit, strong enough to fight a thousand enemies for the thing beloved, with power, with judgment, with endless patience, and to give with everything else a subtler stimulus which is more delicate and perhaps more necessary than all the rest.[10]

This "passion for flowers" and, by implication, for nature is not simply aesthetic or superficial but involves sacrifice. The work of naturalists and nature lovers, of botanists and others is a sacrifice for plants and nature and not a hobby. People ask Celia about the secret to her summer "flower patch" and the winter "window gardens" and she provides her response:

> And I answer with one word, 'Love.' For that includes all, – the patience that endures continual trial, the constancy that makes perseverance possible, the power of foregoing ease of mind and body to minister to the necessities of the thing beloved, and the subtle bond of sympathy which is as important, if not more so, than all the rest. For though I cannot go so far as a witty friend of mine, who says that when he goes out to sit in the shade on his piazza, his Wistaria vine leans toward him and lays her head on his shoulder, I am fully and intensely aware that plants are conscious of love and respond to it as they do to nothing else.[11]

Celia explains that love is the secret and tells the story of a friend of hers who sees the love between people and plants as being mutual. Moreover, Celia likens nature to art:

> Like the musician, the painter, the poet, and the rest, the true lover of flowers is born, not made. And he is born to happiness in this vale of tears, to a certain amount of the purest joy that earth can give her children, joy that is tranquil, innocent, uplifting, unfailing.[12]

Like the artist and others, the "lover" of flowers finds joy amid sorrow. Celia is poetic, philosophical and practical about the matter of flowers and nature.

Pfister also stresses some of the significant contributions of the other members of the Thaxter family:

> As for Levi, he collected birds, that is he shot them as specimens. He took over the education of Roland and John when Celia (more or less)

left him. That education was deep in nature studies and in Latin and Greek. Roland's artistic work shows most clearly in his sketchbook that is available online. His Laboulbeniales, and other illustrations, are remarkable for their clarity and meticulous detail.[13]

Before getting to some of my general observations on the Thaxters, I will briefly address Pfister's observation on Roland Thaxter's illustrations. Roland's sketchbook gives some sense of him at work on his drawings and is available online.[14] Like his parents, Roland combines strong powers of observation with a sense of art and nature. He has a gift for words and making images. The Farlow Reference Library and Herbarium of Cryptogamic Botany at Harvard includes the specimens and original drawings of Roland Thaxter. The library record states that Mabel Gray Thaxter (1858–1952), Roland's wife, gave the Thaxter collection to the Farlow Reference Library.[15] Roland Thaxter has a sense of style in his writing as can be seen in the opening of one of his works:

> DURING a stay of several months in Chile, which extended from November, 1905 to March, 1906, the writer paid but scant attention to the collection of Laboulbeniales; since, in the absence of any very considerable number of insect hosts, his time and attention was largely occupied with the very rich cryptogamic flora of this region, large collections of which were made. The comparative rarity of hosts, especially in the heavily forested area about Corral, was surprising; and, with few exceptions, the parasites which occurred on them were not characterized by any unusual peculiarities.[16]

The syntax, rhythm and clarity also represent something dramatic, which can also be seen in the diction – "rarity" and "surprising." Roland then shifts from this surprising rarity to parasites that did not show "any unusual peculiarities."

Roland had a literary and artistic education as well as a scientific one, including his studies with William Gilson Farlow (1844–1919). The Thaxter Family had long connections to Harvard, Boston and New England. Levi Lincoln Thaxter (1824–1884), a Harvard graduate, had an interest in literature and law. He married Celia Laighton Thaxter (1835–1894), who became one of the best-known poets in the United States. Their son Roland (1858–1932), also a Harvard graduate, was a leading scientist at Harvard. This family, like the Darwins and Huxleys in England, combined arts and science, and their story deserves to be better known in this generation. The family at Harvard and in Boston and New England and their web of friends and colleagues in these communities and in the wider world represent the world of culture and nature. These three members of the

Thaxter family shared a love of nature, art, language and poetry and represent worldviews that are interdisciplinary in which male and female explore the diversity of nature and culture through science and art, image and word. Levi read Shakespeare and championed Robert Browning in Boston where he offered public readings. Celia wrote poetry and kept a garden on the Isle of Shoals and collected seaweed which she mounted and arranged artistically. She painted and was commissioned to paint china, often using botanical motifies. Celia's brother, Oscar Laighton, who lived to almost 100, wrote poetry, prose and drew ships.[17] William H. Weston remembers Roland Thaxter: "Although austere and reserved, Dr. Thaxter had a keen appreciation of beauty which was manifest in his love of music, art, and literature."[18]

Roland collected samples of microfungi on his trips, kept a journal and produced highly elaborate illustrations for his taxonomic work. Celia is a major figure who is an instance of a woman expressing herself and living a rich and independent life. This family is an example to us now of people being accomplished throughout their lives.

In the Thaxter story, told briefly here, I am calling attention to some details. Levi Thaxter and Thomas Wentworth Higginson, friends and correspondents, made important contributions to spreading the reputation of Browning in Boston and the United States.[19] Alexandra Sutherland Orr published the letter of 14 March 1880 from Celia Thaxter to Robert Browning in which she responds to his note and in which she tells Browning that Levi knows "every word you have ever written" and Roland wrote Browning for a poem for an epitaph for Levi's grave.[20] Higginson was one of the admirers of Celia's work on the sea and her descriptions of the coast of the Isle of Shoals. Levi and Celia Thaxter shared a love of poetry so I would like to amplify on their connection with Robert Browning to which I alluded to above. Orr gives a good account of an exchange: "The names of Levi and Celia Thaxter of Boston had long, I believe, been conspicuous in the higher ranks of his disciples, though they first occur in his correspondence at about this date."[21] To Browning, Celia wrote a letter from Newtonville on 14 March 1880, which opens:

> Your note reached me this morning, but it belonged to my husband, for it was he who wrote to you; so I gave it to him, glad to put into his hands so precious a piece of manuscript, for he has for you and all your work an enthusiastic appreciation such as is seldom found on this planet: it is not possible that the admiration of one mortal for another can exceed his feeling for you.[22]

Celia acknowledges receipt of Browning's note and that she passed it on to Levi and she then describes his enthusiasm for Browning in the most superlative terms. Among the details of her description, Celia says that Levi "knows every word you have ever written" and adds: "He reads you aloud (and his reading is a fine art) to crowds of astonished people, he swears by you, he thinks no one save Shakspere has a right to be mentioned in the same century with you. You are the great enthusiasm of his life."[23] For Levi, law, poetry and science were central interests, and Celia stresses his love of Browning's poetry. In the last paragraph of the letter that follows, Celia asks Browning's pardon for smiling because, undoubtedly, he hears many of these things. Then Celia, ever close to the sea, creates an image of the ocean: "But a genuine living appreciation is always worth having in this old world, it is like a strong fresh breeze from off the brine, that puts a sense of life and power into a man. You cannot be the worse for it."[24] Celia likens appreciation to a wind off the salt water that invigorates and empowers a person. Her description of Levi's appreciation leads to a direct address to Browning ("You") and a use of litotes or understatement that means he will be the better for it. Celia blends the bold direct address with the more ironic and modest litotes.

Orr also writes that after Levi died in February 1885, Roland wrote to Robert Browning, asking him for some lines to be put in his father's tombstone. Orr thinks this is the first instance of this poem's publication and she reproduces it:

> *Written to be inscribed on the gravestone of Levi Thaxter.*
> Thou, whom these eyes saw never, say friends true
> Who say my soul, helped onward by my song,
> Though all unwittingly, has helped thee too?
> I gave but of the little that I knew:
> How were the gift requited, while along
> Life's path I pace, could'st thou make weakness strong,
> Help me with knowledge for Life's old, Death's new!
>
> R. B.
> April 19, '85.[25]

Browning's poem is to a friend he never met in person, whose friendship was a mutual help, the requited gift, a request that the addressee – Levi – make strong a weakness and help the speaker – Browning – with the knowledge of the old of life and the new of death. The speaker, alive, speaks to someone recently alive but now dead, an invocation or apostrophe to the dead friend to help in their requited friendship, which was never embodied in person. Levi's father, also Levi

Thaxter, also had poetry on his tomb, buried in Lot 409, Linden Path, Mount Auburn Cemetery, the verse on the left being the first stanza of the hymn "No bitter tears for thee be shed" by Felicia D. B. Hemans and the verse on the right is the third stanza of her poem "Dirge of a Child."[26] Verses on the tomb are a convention of the time and a tradition of the family. This is a family that appreciates and expresses itself through poetry.

Celia's poetry shows sensitivity to the natural world, to plants on land and sea. She is aware of the beauty and the power of nature. One instance is the opening stanza of "Contrast," which begins her collection of poems, *Drift-wood*:

> The day is bitter. Through the hollow sky-
> Rolls the clear sun, inexorably bright,
> Glares on the shrinking earth,- a lidless eye,
> Shedding no warmth, but floods of blinding light.[27]

Amidst the personification of "the hollow sky" and the sun as "a lidless eye," Celia Thaxter creates an opening landscape of a bitter day "of blinding light." The poet and speaker observe nature and feel it, and Thaxter intensifies that experience through adjectives (as present participles "shrinking," "Shedding" and "blinding") and the adverb "inexorably" in addition to the verbs "Rolls" and "Glares" to begin the second and third lines. The image is of a sun that is "clear" and "bright" that blinds with its "light," which the rhyme reinforces. The rhyme also stresses a "sky" that is hollow and the sun an "eye" without lids. The day is without warmth and is bitter. The speaker is in and of nature and observes and feels it and translates that situation into words.

Celia wrote in various kinds or genres. There are many letters of Celia in Houghton Library at Harvard. For instance, she writes to Annie (Adams) Fields in 1875 about nature in the mountains in Vermont: "& it is my first thorough experience of mountains – autumn foliage, – I have learned what incandescence means, only looking at the marvellous maple trees. Such garnet, ruby, toper, amber, gold a flame!"[28] Celia also writes to Ralph Waldo Emerson and William Dean Howells. On 28 November 1869, she addresses Emerson: "And let me say I heard your lecture on 'Poetry and Criticism' and felt like a dried leaf in the autumn wind as I listened to your searching words."[29] In an undated letter probably from 1875 and one on 17 April 1879, Celia praises Howells for the beauty and pleasure of his work.[30]

One of Levi's classmates at Harvard, William Morris Hunt (31 March 1824–8 September 1879), the renowned painter, Celia discovered drowned on the Isle of Shoals. Houghton has a manuscript in which, years after, Celia describes the last days and death of Hunt. Celia writes:

> He took great pleasure in my little garden, - ~~that dear little garden I William [p. 13]Hunt will always be associated with it in my mind, he enjoyed it so much.~~ It was a wilderness of blossoms that summer, there were hollyhocks and sunflowers, clouds of pink, blue and purple cornflowers, golden coreopsis, fiery marigolds, nasturtiums, scarlet beans, the brightest, most delicate sweetpeas, mignonette, poppies, larkspur, white phlox, pansies, all sorts of lovely things. He marvelled with me at the wondrous blue of the cornflowers. 'It is the most perfect blue in the world,' he said, 'there is nothing so beautiful, no blue that approaches it.'
>
> Appleton Brown often sat in the east end of the garden among the flowers, painting them, with their exquisite background of seas and sails. One day Hunt took his brush from the younger painter's hand and began to paint himself, rapidly, feverishly, and worked an hour on the picture. He was very much excited and was worse after it for several days. Brown kept the picture as a most precious possession, the last thing that Hunt ever painted in this world.[31]

This manuscript and pictures from books about Hunt and a painting from the Harvard Portrait Collections help to tell this story. A student of Hunt, Helen M. Knowlton, wrote about him, someone Levi and Celia had long known, in two books.[32] The themes of arts and science, poetry and biology, culture and nature are those in my book and the proposed exhibition that Donald Pfister and I were proposing in 2019. One of the most germane of Hunt's portraits is that of Jean Louis Rodolphe Agassiz, who studied with Humboldt and Cuvier in Paris and was a professor of natural history in Switzerland before becoming a professor of geology and zoology at the Lawrence School at Harvard and the founder of the Museum of Comparative Zoology at Harvard.[33] Donald H. Pfister provides some helpful background: "I believe there are two at the Art Museum. I also have a recollection that there is a portrait of Levi by Hunt now in the hands of Jonathan Hubbard a great great grandson" and adds: "One of Hunt's paintings the Bathers was once described to me as using Roland as a young boy as a model."[34] At Harvard, the world of art and science was and is intertwined.

Celia corresponded with Longfellow,[35] who wanted to know the background to her poem, "The Spaniards Graves," and with Whittier.[36] Annie Fields (Mrs. James T. Fields on the title page), who wrote *Whittier: Notes of His Life and His Friendships* (New York: Harper and Brothers, 1893), seems to be a friend in common with Celia Thaxter, Longfellow and Whittier. On 13 March 1887, Celia writes Whittier: "You ask if Mohini is as remarkable in conversation as in appearance... Shakespeare is as familiar to him as to the greatest Shakespeare student among us & he is equally at home in the domain of science & philosophy."[37] Mohini, Protap Chunder Mozoomdar (1840–1905) author of *The Oriental Christ* (Boston: Geo. H. Ellis, 1883), was lecturing in Boston in the fall of 1883 in pulpits and at the Lowell Institute. Celia also writes to Whittier from the Isle of Shoals on two occasions in letters that are in Houghton, the first, on 24 June 1888 – "My garden is full of golden flowers in memory of you" and the second on 11 April 1889: "Every year I plant the garden for your dear eyes, with yellow flowers. I never forget those lovely summers long ago when you came & loved my flowers."[38] In a beautiful work of poetry, *Snow-Bound. A Winter Idyl* (1866), Whittier creates atmosphere about season and nature, observing, for instance: "In tiny spherule traced with lines / Of Nature's geometric signs."[39] Whittier, a Quaker and an abolitionist, was a friend to Oliver Wendell Holmes, Celia Thaxter, Annie Fields, Sarah Orne Jewett and others and a contributor to *The Atlantic Monthly* (whose editor was James Russell Lowell, who seems to have been married to Levi Thaxter's cousin Maria White, whom Levi mentions in letters to Wentworth Higginson). Whittier wrote about his family farm in Essex county that had been in the family since the seventeenth century. He sent Celia a poem and she wrote to him (in a letter in Houghton):

> How I wish I could see you in this warm greenery of mine, where all the four deep window seats are full of ferns, & flowers that blossom like mad! It is so pleasant a winter parlor, full of sunshine & green leaves & blossoms. It seems very long since I saw you. But I think of you very, very often & love you most tenderly & faithfully, & wish for you every beautiful thing for every moment of all your existence & am yours always Celia Thaxter.[40]

Celia Thaxter also had a sense of the natural world that combined art and science. A beautiful work to display is her seaweed gathered on the Isle of Shoals. Here nature is the artist. As a poet who has worked with artists on artist books, I have long wished to make an artist book by writing poetry in response to the images of Celia's seaweed, which have no words with them.[41] This was to be part of the exhibition that did not come to fruition. Donald H. Pfister and I had hoped that the images from this album would be reproduced to make an artist book. This book and exhibit both in person and digitally could have introduced the art and science of the Thaxter family to members of the Harvard and wider communities. Both the book and exhibition were to be a meeting of past and present, science and art, nature and culture. In part, this chapter attempts to convey some of the dimensions of the Thaxters, a prologue, hope to my further research and writing on them and prompt to further research by others.

Celia has many letters, not all of them in Houghton, but in that library there is also a copy of *Letters of Celia Thaxter. Edited by her Friends A. F. and R. L.* Boston and New York: Houghton, Mifflin and Company. The Riverside Press, Cambridge, 1895, which Sarah O. Jewett signed in 1895.[42] In another copy appears a note from Celia's son, Roland:

> Harvard University
> Cambridge
> 3 Scott St. June 5. '95
> Mr. W. H. Tillinhast
> Cambridge
>
> Dear Sir:
>
> The letters of my mother were edited by Mrs James T. Fields Fields and Miss Rose Lamb, the former being of course the A. F. of the letters.
> In haste truly yours
> Roland Thaxter:[43]

Here, Annie Fields appears again. Roland himself was, as discussed before, a key figure at Harvard and the United States in science.

Roland Thaxter was for a period a student in the medical school at Harvard and then a student of cryptogamic botany under William Gilson Farlow (himself gifted in music and science and worthy of further research) in the Graduate Faculty of Arts and Science; the Farlow Library and Herbarium has some important items relating to Roland's writings and interests in biology.[44] A distinguished mycologist, Roland was accomplished in many groups of fungi but is best known for his study of Laboulbeniales, an order of fungi associated with insects. His five-volume illustrated series (1896–1931) is a tour de force in descriptive mycology. *The Harvard Crimson* reported Roland's death:

> Roland Thaxter '82, professor of Cryptogamic Botany, Emeritus, and Honorary Curator of the Farlow Herbarium of Cryptogamic Botany, died yesterday at his home. He was seventy-three years of age and had been ill for several months. Professor Thaxter was born in Newton, August 28, 1858, and was the son of Levi L. and Celia (Laighton) Thaxter. He was graduated from Harvard with the degree of A.B. in 1882, and received his A.M. and Ph.D. in 1888.[45]

After Roland Thaxter died, other important pieces appeared on him. One observed:

> His first researches were on the rust *Gymnosporangium*, but he published an account of the entomogenous family Entomopthoreæ in 1888, which still remains one of the most important papers on the group. A few phytopathological reports appeared while he was at Connecticut, some of considerable interest. In 1890 he made his first contribution to our knowledge of the peculiar entomogenous fungi, the Laboulbeniaceæ, and it is with these that his name is chiefly associated, for from being almost unknown, they became, owing to Thaxter's enthusiasm, a group as well understood as any other and containing hundreds of genera and thousands of species.[46]

In his biographical memoir, G. P. Clinton sees Roland Thaxter as a scientist, artist and poet and he places him not only in the context of his parents Levi and Celia but also his ancestors on both sides, including a grandfather and great-grandfather who went to Harvard. Clinton mentions Roland's many trips to collect fungi from Kittery Point (where Roland and his father lived and are buried and in Maine close to the Isle of Shoals) through Jamaica to South America, where in 1905–1906 in Chile he collected and that led to "Notes on Chilean Fungi" in the *Botanical Gazette* in December 1910.[47] Clinton's biographical memoir has a picture of Roland.

This chapter has focused briefly on the Thaxters, Levi, Celia and Roland, a New England and Harvard family accomplished in arts, letters and science with connections in Boston, New England, the United States and beyond. Celia is also important for gender as she was a leading poet in the United States and all of them have connections to abolitionists, artists, writers and scientists that tell a story important then and now. The seaweed album of Celia's is also remarkable. As Donald H. Pfister notes:

> these seaweed albums were a popular activity for women of this era. We [the Farlow Library and Herbarium] have a number of them. One by Mrs. Robinson is a delight. ... All this is to say that Celia was not unique in this activity.[48]

Celia brings together art, culture, literature and science connecting us to land and sea in New England and beyond. The Thaxters' love of nature informed their lives and their art and science. Their ecology can inform ours in this and other places, at Harvard and in the wider world.

In the next section, I turn to Marie Curie and the Curie family, who made such important contributions to physics, chemistry and medicine as well as being committed to languages, humanities and the arts, and to Bertrand Russell, another figure from a remarkable family who combined science with art, who was a writer as well as a logician, philosopher and mathematician. Nature, literature and culture are not readily separated even in a specialized world. It seems to me that in the arts and humanities, evidence, testing and interpretation are keys as well as in science, what was once called natural philosophy.

Notes

1 Celia Thaxter, *An Island Garden with Pictures and Illuminations by Childe Hassam* (Boston and New York: Houghton, Mifflin & Co, 1894. The Riverside Press, Cambridge MDCCCXCV), [np]. For a digital copy, see https://digital.library.upenn.edu/women/thaxter/garden/garden.html.

2 THAXTER, *An Island Garden*, [np].

3 See, for instance, Donald H. Pfister, "A Note on Roland Thaxter's Collection Localities," *Mycotaxon* 15 (1982): 341–344; Donald H. Pfister, "The W. G. Farlow Correspondence," *Taxon* 31 (1982): 607–608; Donald H. Pfister, "Roland Thaxter: Mycologist and Entomologist," [Chapter 9] in *Fungus – Insect Relationships – Perspectives in Ecology and Evolution*, edited by Q. Wheeler and M. Blackwell (New York: Columbia University Press, 1984), 211–221; Donald H. Pfister, "An Index to the Taxa Described by Roland Thaxter Excluding the Laboulbeniales," *Mycotaxon* 20 (1984): 225–244; Donald H. Pfister, "Roland Thaxter and the Myxobacteria," [Chapter 1] in *Myxobacteria II*, edited by M. Dworkin and D. Kaiser (Washington, DC: Published by the American Society for Microbiology, 1993), 1–12.

4 My thanks to Donald H. Pfister for his encouragement, for his permission to discuss our proposed exhibition or display here and for his comments then and on drafts of this chapter.

5 Comments on my discussion in the file on the Thaxters in Donald H. Pfister to Jonathan Locke Hart, electronic mail, 6 November 2025.

6 Donald H. Pfister to Jonathan Locke Hart, electronic mail, 6 November 2025. See also Celia Thaxter, *By This Wing : Letters by Celia Thaxter to Bradford Torrey about Birds at the Isles of Shoals, 1888 to 1884* (1st. ed.) (Manchester: J. Palmer Publisher, 1999).

7 See https://www.metmuseum.org/art/collection/search/14930

- Title: Celia Thaxter's Garden, Isles of Shoals, Maine
- Artist: Childe Hassam (American, Dorchester, Massachusetts 1859–1935 East Hampton, New York)
- Date: 1890
- Culture: American
- Medium: Oil on canvas
- Dimensions: 17 3/4 × 21 1/2 in. (45.1 × 54.6 cm)
- Credit Line: Anonymous Gift, 1994
- Object Number: 1994.567
- Curatorial Department: The American Wing.

8 Celia Thaxter, *An Island Garden*, 3.

9 See Henry David Thoreau, *Faith in a Seed : The Dispersion of Seeds and Other Late Natural History Writings* (Washington, D C: Island Press/Shearwater Books, 1993).
10 Celia Thaxter, *An Island Garden*, 4.
11 Celia Thaxter, *An Island Garden*, 5.
12 Celia Thaxter, *An Island Garden*, 5–6.
13 Donald H. Pfister to Jonathan Locke Hart, electronic mail, 6 November 2025.
14 For Roland Thaxter's Sketchbook in manuscript, see **Persistent Link** https://nrs.lib.harvard.edu/urn-3:fmus.gra:28700537

Description: Farlow Reference Library and Herbarium; Untitled sketchbook; Botany Libraries, Archives of the Farlow Herbarium of Cryptogamic Botany, Harvard University Herbaria, Cambridge, Mass.
Repository: The Gray Herbarium Library
Institution: Harvard University
Accessed: 13 November 2025

See also Meredith Blackwell, Danny Haelewaters and Donald H. Pfister, "Laboulbeniomycetes: Evolution, Natural History, and Thaxter's Final Word," *Mycologia* 112 (6) (2020): 1048–1059.
15 See https://hollisarchives.lib.harvard.edu/repositories/20/resources/7516.
16 Roland Thaxter, "New Libocibentales from Chile and New Zealand," *Proceedings of the American Academy of Arts and Sciences*. Vol. 54. No. 3 September, 1918. Contributions from the Cryptogamic Laboratory of Harvard University. No. LXXXII.
17 See https://www.brayco.com/auction-lot/two-ship-drawings-by-oscar-laighton-with-original_45045BC866.
18 William. H. Weston, "Roland Thaxter," *Mycologia* 25 (2) (1933): 69–89.
19 For this point, see Hédi Jaouad, *Browningmania, America's Love for Robert Browning* (Amherst, NY: Cambria Press, 2014); for letters from Levi to Wentworth, see the Higginson papers at Houghton, bMS Am 1162.10 (920–926), Thaxter, Levi Lincoln, d. 1884.including Levi's interest in Shakespeare at A.L.S.; same, 26 January 1842. 1s, (4 p.).
20 Jaouad cites this letter – Orr, *Life and Letters*, 351–352, see Jaouad.
21 Mrs Sutherland Orr, *Life and Letters of Robert Browning* (London: Smith, Elder, & Co., 1891), 351.
22 Orr, *Life*, 351–352.
23 Orr, *Life*, 352.
24 Orr, *Life*, 352.
25 Orr, *Life*, 353.
26 For the hymn, see https://hymnary.org/text/no_bitter_tears_for_thee_be_shed#Author and for the poem, see Felicia Hemens, *Poems of Felicia Hemens* (Edinburgh: William P. Nimmo, 1875), 54. My thanks to Donald H. Pfister for calling my attention to this verse and the tome; Donald H. Pfister to Jonathan Locke Hart, electronic mail, 14–15 November 2025.
27 Celia Thaxter, *Drift-weed* (Boston, MA: Houghton, Osgood and Company; Cambridge: The Riverside Press, 1879), 1.
28 MS Am 1745.1 (23), Thaxter, Celia (Laighton) 1835–1894. A.L.S. to [Annie (Adams) Fields]; Montpellier Vermont, 19 October 1875. Ls. (2 p.).
29 See bMS AM 1280 (folder 3183), Thaxter, Celia (Laighton), 1835–1894. A.L.S. to RWE; Newtonville, 28 November 1869. 1s. (3p.), The Houghton Library 5158.

30 Howells bMS Am 1784 (478).
31 See Houghton Mifflin Company correspondence and records, 1832–1844, Houghton Mifflin Company 1832.), Thaxter, Celia (Laighton) 1835–1894. The last days of William Morris Hunt. TS.s. with A.MS. revisions; [n.p., n.d.]. 54s.(56p.) Thaxter, Celia (Laighton).
32 Helen M. Knowlton, *Art-Life of William Morris Hunt* (Boston, MA: Little, Brown, and Company, 1899). See, for instance, Hunt's portrait of his mother, painted in Paris in 1850 (see between pages 16 and 27). Helen M. Knowlton's *Hints for Pupils in Drawing and Painting with Illustrations from Charcoal Drawings by William M. Hunt* (Boston, MA: Houghton, Osgood and Company; Cambridge: The Riverside Press, 1879), particularly one of the landscapes at the back of the book.
33 1955.6, https://www.harvardartmuseums.org/collections/object/308480?position=8
34 Donald H. Pfister, comments on the Thaxter section attached to this chapter as part of Donald H. Pfister to Jonathan Locke Hart, electronic mail, 14 November 2025.
35 bMS Am 1340.2 (5482) folder 5482, Thaxter, Celia (Laighton), 1835–1894. 2 letters; 1877–1878.
36 Pickard-Whittier papers, John Greenleaf Whittier, HOU B MS Am 1844 Folder 305.
37 See *The Letters of John Greenleaf Whittier*, Volume 1 (Cambridge, MA: Harvard University Press, 1975), note 6, p. 481 and see also letter 1424 in this volume.
38 See Collection: Pickard-Whittier papers / Series: IV. Other letters. Item –Box 14 Identifier: MS Am 1844 (304, 686, 884, 888).
39 John Greenleaf Whittier, *Snowbound a Winter Idyl* (Boston: Ticknor and Fields, 1866), 11.
40 See Collection: Pickard-Whittier papers / Series: IV. Other letters. Item – Box 14Identifier: MS Am 1844, (304, 686, 884, 888).
41 HOU GEN MS Am 1272, Seaweeds gathered at the Isle of Shoals in 1857. Thaxter, Celia. This would be done in a similar spirit to *Darkfire*, which his co-artists Sean Caulfield, Susan Colberg and I gave to Houghton (Houghton Library GEN HRef-1627).
42 The call number is HOU GEN *AC85.J55554.Zz895t.
43 Found in another version of this book, the letter from Roland Thaxter.
44 Biographical Sketch Source: The Farlow Reference Library of Cryptogamic Botany Archives, The William Gilson Farlow (1844–1919) Collection. See https://fof.huh.harvard.edu/william-g-farlow; on Roland Thaxter, see Harvard University Herbaria & Libraries. https://www.huh.harvard.edu/book/roland-thaxter.
45 "Thaxter, Professor of Botany, Dies at Home," *Fellow of American Academy of Sciences*, No Writer Attributed April 23, 1932, The Harvard Crimson; https://www.thecrimson.com/article/1932/4/23/thaxter-professor-of-botany-dies-at/.
46 *Nature*, Volume 130, pp. 84–85 (1932).
47 George Perkins Clinton, "Biographical Memoir of Roland Thaxter, 1858–1932", *National Academy of Sciences*, 17 (1935): 55–68.
48 Donald H. Pfister comments on the Thaxter section attached to this chapter as part of Donald H. Pfister to Jonathan Locke Hart, electronic mail, 6 November 2025.

6 Marie Curie and Representing Pierre Curie

The Early Years

Having examined the role of family and representation in science regarding Charles Darwin, Erasmus Darwin, Asa Gray, Jane Loring Gray and the Thaxters, I continue in discussing Marie Curie and her portrayal of her husband Pierre Curie and herself in this chapter and the next chapters. How does reflection on one's own life or the life of another, a scientist in one's family, affect the representation of science? The nature of natural philosophy, what we came to call science, was and is beneficial. A scientific way of thinking or proceeding to examine phenomena and problems is good in culture and humanities in theory and practice. Evidence, facts, examples and the attempt at general laws and the testing of these with evidence are, for me at least, central to science and culture. As we go along, we adjust and correct according to the facts and evidence. Otherwise, we just opine, and assert. Humanities and the arts are not the same as science but share qualities with science. So, by examining those working in biology, geology, logic and mathematics, as well as discussing those who worked in history, literary theory, literature and poetry, I am trying to show the overlap. My method of close reading, paying close and detailed attention to texts, in science and culture is what, as I mentioned in the Introduction, underlies this book, is its primary focus. Just as Darwin pays attention to words and texts, so does Marie Curie. Her biography of Pierre Curie and her autobiography are the concern of the chapters on the Curies, especially as Marie represents matters in relation to science and scientific practice, theory and worldview.

The books I will discuss are the English and French versions of Marie Curie's biography of Pierre Curie, including the English version which includes an autobiography that, she says, some American friends requested. The French version does not contain the autobiography and because I am writing here in English, and not in French as I sometimes do, I will stick to the English version of the autobiography, while noting that this is a translation, which alerts us to what is lost and gained in translation.[1] To reiterate, the English work makes sense as it makes readily available the

DOI: 10.4324/9781003651871-7

autobiography and is accessible, as this is a book in English. Although I quote from the French to enhance the analysis and remind the reader of the original language, increasingly, I came to translate the original because the English translation does not follow the original as closely as I might like for the scholarly purposes of my readers. Thus, I recommend the translation by the Kelloggs, but in performing close reading and analysis of the biography of Pierre Curie, I often translate. As for the autobiography of Marie Curie, I try to stick to the English translation of the Kelloggs owing to its availability in one book and because, as Marie suggests, the request for it came from the United States. Thus, I try to balance availability, readability for those who read in English and not in French and accuracy or being close to Marie Curie's syntax and diction.

The epigraph of the book on Pierre Curie begins with his words at the Nobel conference of 1903:

> On peut concevoir encore que dans des mains criminelles le radium puisse devenir très dangereux, et ici on peut se demander si l'humanité a avantage à connaître les secrets de la nature, si elle est mûre pour en profiter ou si cette connaissance ne lui sera pas nuisible.[2]

The Kellogg translation that appeared in 1923 renders the sentence thus:

> It is possible to conceive that in criminal hands radium might prove very dangerous, and the question therefore arises whether it be to the advantage of humanity to know the secrets of nature, whether we be sufficiently mature to profit by them, or whether that knowledge may not prove harmful.[3]

The translation of "On peut" as "One can" or "We can" or, as the Charlotte and Vernon Kellogg have it, "It is possible," are all feasible, and they have the first one way, as discussed, and the second "on peut" as "we be." I might make the translation more personal by repeating the "we" construction:

> We can conceive still that in criminal hands radium could become very dangerous, and here we can ask ourselves whether humanity has an advantage in knowing the secrets of nature, whether it is ready to profit from them or whether this knowledge will not be harmful to it.

Besides trying to reproduce Pierre Curie's style (his diction and syntax), I use this example to foreground translation as part of the challenge of close reading or analysis, especially in different languages. As I have a chapter for my analysis of the writing of Marie Curie (after this brief discussion of a sentence she quotes from Pierre as the opening of an epigraph), I cannot perform an extensive analysis of the original text and its relation to the original without expanding it into a book. Marie uses Pierre's words to frame the book she is writing about him. He muses whether humans are mature enough to know and use radium as one of "the secrets of nature" for good and not for ill.

Marie begins her Preface with a disclaimer: "It is not without hesitation that I have undertaken to write the biography of Pierre Curie"[4] ("Ce n'est pas sans hésitation que j'ai accepté d'écrire la biographie de Pierre Curie").[5] She explains why she hesitated and her preference: "I should have preferred confiding this task to some relative or some friend of his infancy who had followed his whole life intimately and possessed as full a knowledge of his earliest years as of those after his marriage"[6] ("J'aurais préféré confier cette tâche à quelque parent ou ami d'enfance, qui eût connu de près son existence, aussi bien dans ses plus jeunes années que par la suite").[7] Pierre's brother Jacques, a fellow scientist, had moved to University of Montpellier and thus "insisted that I should write the biography, believing that no one else better knew and understood the life of his brother"[8] ("il insista donc pour que j'écrive la biographie, pensant que nul ne pouvait mieux que moi connaître et comprendre la vie de son frère").[9] Marie notes what strategy she chose under the circumstances. Of Jacques, she says: "He communicated to me all his personal memories; and to this important contribution, which I have utilized in full, I have added details related by my husband himself and a few of his friends"[10] ("Il m'a communiqué tous les souvenirs personnels qu'il a pu réunir. A cette contribution importante, que j'ai utilisée intégralement, j'ai joint des details qui m'ont été racontés par mon mari et par quelques amis de celui-ci." The Kelloggs leave out "qu'il a pu réunir" ("that he was able to gather"), but other than that, the translation follows the original closely. Marie does her best to construct her narrative, especially of Pierre's life before she met him:

> Thus I have reconstituted as best I could that part of his existence that I did not know directly. I have, in addition, tried faithfully to express the profound impression his personality made upon me during the years of our life together[11] (J'ai ainsi reconstitué de mon mieux la partie de son existence qui ne m'est pas dïrectement connue. J'ai, d'autre part, essayé de traduire fidèlement la profonde impression que j'ai eue de sa personnalité pendant les années de notre existence commune).[12]

The translation is good and close to the original with one change in the syntax, which would be "I thus reconstituted," and it uses "faithfully to express" when literally it would be "translate faithfully" while translating "d'autre part" as "in addition" when it usually signifies "on the other hand" and while rendering "notre existence commune" as "our life together" when a more literal translation would be "our common existence" or "existence in common," which would be closer to Marie Curie's diction, as she uses "l'existence" twice in the sentence rather than "la vie." The Kelloggs may be thinking of what sounds better in English as opposed to showing the word choice and music of Marie Curie's prose. Curie stresses she is doing her best and being as faithful as she can be. Marie ends the Preface with a touching and personal paragraph:

> This narrative is, to be sure, neither complete nor perfect. I hope, nevertheless, that the picture it gives of Pierre Curie is not deformed, and that it will help to conserve his memory. I wish, too, that it might remind those who knew him of the reasons for which they loved him[13] (Ce récit n'est certes ni complet, ni parfait. J'espère cependant que l'image qu'il donne de Pierre Curie n'est point déformée et qu'elle aidera à conserver sa mémoire. Je souhaite aussi qu›elle évoque, pour ceux qui l›ont connu, les raisons pour lesquelles ils l›ont aimé.).[14]

Curie's "l'image" can be rendered as "image" or as "picture" as the Kelloggs translate it. Marie stresses incompletion, imperfection and a hope not to deform the image of Pierre. She wishes the book to provide a conservation of his memory and a reminder, to those to whom Pierre was known, of why "they loved him." The translation of the French is very close, for she says: "Je souhaite aussi qu'elle évoque, pour ceux qui l'ont connu, les raisons pour lesquelles ils l'ont aimé."[15] Her original French shows that she wishes that his memory will evoke, for those who knew Pierre, the reasons for which they loved him. This book is written out of love for Pierre Curie. In the English version, she signs this prefatory matter M. C. and the French edition Mme CURIE, the one the initials of her name, the other the wife and widow of Pierre Curie.

Marie Curie speaks about the origins of the Curie family in Alsace, Protestants, and a family given to medicine and science. Pierre's father had an interest in science and laboratories: "Bien que son père fût établi à Londres, Eugène Curie avait été élevé à Paris où il fit ses études de sciences naturelles et de médecine, et travailla comme préparateur dans les laboratoires du Muséum, auprès de Gratiolet"[16] ("Although his father was based in London, Eugène Currie was raised in Paris, where he studied natural sciences and medicine and worked as a laboratory technician [preparator] in the laboratories of the Museum, working with Gratiolet," my translation). For laboratory technicians, the Kelloggs use "preparator," which is close to the French but perhaps a rarer word; the *Oxford English Dictionary* (OED), which says the word originated from about 1485 and was rare before the nineteenth century, defines it as "A person who or (occasionally) thing which prepares something; a preparer of medicine, specimens, exhibits, etc."[17] Marie Curie notes that Pierre's father Eugène Curie was a doctor and the son of a doctor and that even though "he wished to consecrate his life to scientific work, family responsibilities following his marriage and the birth of two sons, forced him to renounce this desire." The French is less concessive in its structure: "Ayant souhaité consacrer sa vie au travail scientifique, il dut renoncer à ce projet en raison de charges de famille que lui imposèrent son marriage et plus tard la naissance de deux fils."[18] I would translate this more literally for those who do not have French as "Having wished to consecrate his life to scientific work, he had to renounce this project for the reason of charges [expenses, responsibilities] of the family imposed on him by his marriage and later the birth of two sons." Family circumstances affect science.

As in Darwin's narrative, in Marie Curie's account there is an exploration of a family with medical and scientific interests. Marie notes that necessities obliged Eugène Curie to become a doctor, but he did some experimental research as his means permitted, for instance, investigating inoculation for tuberculosis when it was not established at a time when the bacterial nature of this malady was not yet established. She adds:

> Jusqu'à la fin de sa vie, il conserva le culte de la science, et sans doute aussi le regret de n'avoir pu s'y consacrer uniquement. Les préoccupations scientifiques du docteur Curie lui avaient donné l'habitude d'excursions, à la recherche de plantes et d'animaux nécessaires & ses experiences son amour de la nature entretenait d'ailleurs chez lui une préférence marquee pour la vie à la campagne.[19]
>
> Until the end of his life, he maintained a worship [cult] of science, and no doubt also the regret of not having been able to devote himself solely to it. Dr. Curie's scientific concerns had given him the habit of excursions, in search of necessary plants and animals for his experiments, and along with his love of nature maintained in him a marked preference for country life.
>
> (my translation)

Here, Marie Curie finds a tradition of experimental science between Pierre and his father. There is science in the family. The implication is that this milieu gave Pierre encouragement to pursue that endeavour. She talks about Eugène's dedication and selflessness and the medal of honour the French republic awarded him for treating the wounded in the revolution of 1848.

Marie gives many details of the families of Pierre's father and mother, noting that Pierre never left the family for school, not having completed studies at a lycée or high school (secondary school). She says: "Les qualités intellectuelles de Pierre Curie n'étaient point decelles qui permettent d'assimiler rapidement un programme d'études scolaires"[20] ("Pierre Curie's intellectual qualities were not those that would allow him to assimilate quickly a school curriculum"). Marie elaborates in her representation of Pierre. She considers that he had the spirit of a dreamer and would not submit to the regulation of school. Others and Pierre included thought he was slow-witted because he would not follow the regimen. Marie says that she believes this view was justified because she thinks that from an early age,

> his mental faculties compelled him to concentrate his thoughts on a specific object with a fairly great intensity, until he obtained a precise result, without it being possible for him to interrupt and modify the course of his reflections according to external circumstances (ses facultés mentales l'obligeaient à concentrer sa pensée sur un objet determine avec une assez grande intensité, d jusqu'à en obtenir un résultat précis sans qu'il lui eût été possible d'interrompre et de modifier le cours de ses réflexions au gré des circonstances extérieures).[21]

Marie sees that this kind of mind has many future possibilities, but that public education has not provided for this intellectual category, which has more representatives than at first glance. I have translated, paraphrased and condensed her words here:

> Il est clair qu'un esprit de cette nature peut contenir en lui de grandes possibilités d'avenir, mais il est non moins évident qu'aucun système d'éducation n'a été prévu par l'école publique pour cette catégorie intellectuelle, qui cependant compte plus de représentants qu'on pourrait le croire à première vue.[22]

As I discussed with Darwin's view of his own mind and those views of it that he represented, the mind of a great scientist, like Darwin and Pierre Curie, is not necessarily conventional or quick. Marie Curie, another accomplished scientist, understands that formal education has a hard time comprehending and encouraging such a mind, a mind given to concentration, detail and singular focus. It seems that Darwin's father and the school that taught him did not understand the power of Darwin's mind. Marie praises the intelligence of Pierre's parents and for their understanding of what their son needed for education, so they avoided deforming his intelligence with dogmas, prejudices, and preconceived ideas or, in the original French: "de ne point peser sur son intelligence pour la deformer selon des dogmes, préjugés ou idées préconçues."[23] The role of family in science, as

I also stressed in regard to the Darwins, Grays and Thaxters, is evident. Marie continues to emphasize this familial bond:

> De cette éducation si libérale, Pierre Curie conserva toujours a ses parents un souvenir reconnaissant. Il grandit en toute liberté, développant son goût pour les sciences naturelles dans des excursions à la campagne, d'où il rapportait des plantes et des animaux pour les expériences de son père. Ces promenades, qu'il faisait soit seul, soit avec les siens, contribuaient à éveiller en lui un grand amour de la nature, et jusqu'à la fin de sa vie il conserva cette passion.[24]

Here is a translation that tries to reproduce the original diction, expression and syntax:

> Of this education, so liberal, Pierre Curie always preserved a memory acknowledging his parents. He grew up in complete freedom, developing his taste for natural sciences on excursions to the countryside, from where he carried back plants and animals for the experiments of his father. These walks, which he took either alone or with his family, helped to awaken in him a great love of nature, and until the end of his life he preserved this passion.[25]

In memory, Pierre acknowledged his parents and their education of him. The freedom they gave him and the experiments his father did helped to encourage his love of nature and natural sciences. He collected and saw how important experimentation was to his father, a doctor given to science. Pierre's parents understood him, gave him scope to be himself and provided him with an experience of nature. Marie observes that children in the city and in classical education have little contact with nature. She says that Pierre's father guided him to learn to observe and interpret facts and to learn about animals and plants in the environs of Paris. In all seasons, he could identify these in forests and fields, streams and ponds. According to Marie Curie, Pierre was especially interested in the ponds, their vegetation, frogs, tritons, salamanders and dragonflies. She also notes that Pierre made great efforts to pursue these interests and would examine animals in his hands. Moreover, Marie says that after she was married to Pierre, they would walk and he would put a frog in her hands and say how pretty it was and how much he liked to make bouquets of wildflowers.[26]

The growth of Pierre's mind and education concerns Marie, and she devotes considerable space to this development. She speaks of the rapid progression of Pierre's knowledge of natural history and his mastery of mathematics while noting the neglect of his classical studies. Through reading, Pierre developed a knowledge of literature and history, his father, a reader who passed on a taste for reading to his son, having a library in French and other languages. When Pierre was about fourteen, A. Bazille, who took care with him and whose son Albert became friends with Pierre, became his teacher and taught him mathematics and also caught him up in Latin.[27] Bazille's teaching influenced Pierre Curie's mind, his awareness of his own scientific capabilities, his remarkable aptitude for mathematics, his distinctive geometric mind and a great ease of vision in space ("un esprit géométrique caractérisé et une grande facilité de vision dans l'espace").[28] Pierre progressed and study became a fascination and joy while being unwavering to his teacher in his gratitude. Marie reports that Pierre told her a detail proving that from then on, he sometimes deviated from a programme of study to pursue a personal investigation. For instance, she says that the theory of determinants captivated him, so that "he realized an analogous construction, but in three dimensions, and applied himself to discovering the properties and uses of these 'cubic determinants'" ("il entreprit de réaliser une construction analogue, mais à trois dimensions, s'appliqua à découvrir les propriétés et l'utilisation de ces 'déterminants cubiques'").[29] Marie observes that at that age, and given his knowledge, he did not have the strength for the undertaking but showed "his nascent spirit of invention" ("son esprit d'invention naissant").[30] Marie Curie amplifies, this time as if speaking in the young Pierre's voice:

> P[l]usieurs années après, préoccupé de réflexions sur la symétrie, il se posait cette question 'Ne pourrait on trouver une méthode générale pour résoudre une équation quelconque ? Tout est une question de symétrie.' Il n'avait pas alors connaissance de la théorie des groupes de Galois qui a permis d'aborder ce problème mais il a été heureux d'en connattre plus tard les résultats, ainsi quel'application géométrique au cas de l'équation du cinquième degré.[31]
>
> Several years later, preoccupied with reflections on symmetry, he asked himself this question: 'Could we not find a general method to solve any equation? It is all a question of symmetry.' He was not then aware of the theory of groups of Galois, which made it possible to address this problem, but he was pleased to learn its results later, as well as its geometric application to the case of the fifth-degree equation.[32]

Marie represents Pierre as asking questions about mathematics even before he had advanced training. He was interested in symmetry and a general method to solve a problem. Pierre did not have the opportunity to study the work of Évariste Galois, who died at twenty, made a contribution to group theory in mathematics, and had addressed the problem. Marie notes Pierre's quick progress in mathematics and physics, which allowed him, at sixteen, to receive his Bachelor of Science degree, what she sees as completion of the hardest stage for him. Henceforth, as Marie says, "il n'avait plus à songer désormais qu'à acquérir des connaissances par un effort personnel et independent dans un domaine de science librement choisi," that is, "he had only to think about acquiring knowledge through personal and independent effort in a freely chosen field of science."[33] Marie has composed a portrait of a young Pierre as someone with powers of concentration and whose parents had been liberal and flexible with his education, someone who was able to get a degree even though formal education cramped his way of thinking and exploring science and who proved he could excel in the degree and then get back to what he prized most – his independence in a field of science he was free to choose. Finally, he could express his nature and his mind, and now with extensive experience and training. Thus, Marie Curie ends the first chapter on the Curie Family as well as Pierre Curie's childhood and first studies.

The second chapter represents the dreams of Pierre's youth, his first scientific works and his discovery of piezoelectricity. Among many details, Marie Curie notes that Pierre Curie was very young when starting his higher education, preparing for a degree in physics, that he attended lectures and did practical work at the Sorbonne, that he worked in a laboratory with his brother Jacques. Furthermore, Marie says that at eighteen Pierre received his degree in physical sciences and that at nineteen, he became a preparator at the Faculty of Sciences in Paris, a position he held for five years, and was responsible for the students doing the physics experiments and where he conducted "his first experimental research" ("ses premières recherches expérimentales").[34] Owing to finances, Pierre had to accept a position as a preparator at the age of nineteen rather than continue his studies for two or three years more. Owing to professional pursuits and research, Pierre could not take courses in higher mathematics and examinations, but, having agreed to work in public education, he was exempted from military service. Marie makes a portrait in words from a photograph:

> L'impression d'une vie intérieure profonde se degage pourtant du jeune visage tel qu'il apparaît sur une bonne photographie du groupe familial composé par le docteur Curie, sa femme et ses deux fils. La tête est appuyée sur la main dans une pose d'abandon et de rêverie, et on ne peut s'empêcher de trouver frappante l'expression des grands yeux limpides & la forme allongée qui semblent suivre quelque vision intérieure.[35]
>
> The impression of a profound inner life emerges, however, from the young face as it appears in a good photograph of the family group composed of Dr. Curie, his wife, and his two sons. The head is resting on his hand in a pose of abandonment and reverie, and one cannot help but find striking the expression of the large, clear eyes & elongated form, which seem to be following some inner vision.[36]

This is part of a family portrait in which Pierre's future wife and now widow looks back and focuses on him, on a youth whose pose suggests unrestraint, dreaminess, inward vision. These are key characteristics that Marie Curie ascribes to Pierre Curie early in the biography. She also stresses the bond he had with his brother Jacques and their work together in the laboratory as well as mentioning other childhood friends and relatives, often with an interest in medicine and science. Through the memories Pierre recounted to her, Mary describes vacations Pierre had, times with Jacques as well as walks of the two and his own solitary walks: "Pierre Curie s'engageait dans des excursions solitaires qui convenaient bien à son esprit méditatif. Il lui arrivait en ce cas d'oublier l'heure et d'atteindre l'extrême limite de ses forces" ("Pierre Curie engaged in solitary excursions that suited his meditative mind well. In such cases, he sometimes forgot the time and reached the very limit of his strength").[37] She elaborates on this portrait, stressing how Pierre contemplated external things rapturously rather than having material difficulties. Another source Marie uses to portray Pierre is his diary, for instance, an entry in 1879 about the countryside, part of which reads: "Oh ! quel bon temps j'ai passé là, dans cette solitude bienfaisante, bien loin des mille petites choses agaçantes qui, à Paris, me mettent au supplice" ("Oh, what a wonderful time I had there, in this blissful solitude, far from the thousand annoying little things that, in Paris, put me to torment").[38] Pierre's entry, as Marie quotes it, records that he did not regret his nights in the woods and his days alone while saying that if he had time, he would recount his musings (daydreams), describing the delicious valley, perfumed with aromatic plants, the foliage, the Bièvre River, a fairy palace with colonnades of hops, the rocky hills red with heather, saying he remembered the Bois de la Minière, his most beloved place, saying he often returned in the evening up the valley "with twenty ideas in my head" ("avec vingt idées en tête").[39] Nature, the plants, all made an impression on Pierre. Marie tries to represent Pierre through a family photograph, his own words, to reconstruct his early life without her.

Marie Curie interprets this entry in Pierre Curie's diary. She says that reflection in the countryside made him happy. In daily life, Pierre suffered and worried owing to obligations and disturbances, and Marie adds that Pierre felt destined for scientific research and needed to understand phenomena to shape a satisfactory theory. Through repetition, Marie emphasizes that futile causes distracted him from reflecting and focusing on a problem and discouraged him. Marie refers to an entry in the diary entitled "A Day Like Many Others," in which he recorded events that kept him from work, concluding that that was his day in which he did nothing and asked why. Marie says that further on Pierre writes about the same subject and takes his title from a celebrated author, which Marie identifies in a note as Victor Hugo's *Le Roi s'amuse*: "Etourdir de grelots l'esprit qui veut penser," which I translate as "To stun with bells the mind that wants to think" and which the Kelloggs render "To deafen with little bells the spirit that would think."[40] Marie then quotes at length from this entry in Pierre's diary, and here I will only quote some of it to give a sense of his thinking:

> Pour que, faible comme je le suis, je ne laisse pas ma tête aller à tous les vents, cédant au moindre souffle qu'elle rencontre, il faudrait que tout fût immobile autour de moi ou que lancé comme une toupie qui, ronfle, le movement même me rende insensible aux choses extérieures.

which I render as

> So that, weak as I am, I do not let my head go to all the winds, yielding to the slightest breath it encounters, it would have to be that all would be still around me or thrown like a top that, snoring [whirring?], the very movement would make me insensitive to external things.[41]

Pierre develops this image of the top. He is poetic, literary, philosophical. He uses metaphors, similes, images, analogies to make his point about focus and distraction. In likening himself by implication to a top, Pierre picks up on "lancer" to throw or toss or launch and the movement of turning or spinning. Pierre talks about spinning or turning slowly around him, and when he tries to launch himself or take off, something stops him, sets him back or delays him for ever from concentrating and not succumbing to the sweetest things in life, as if to use "anti-natural thoughts" ("les pensées anti-naturelles") to resist nature.[42] He ends the passage Marie has selected with "il faut faire de la vie un rêve et faire d'un rêve une réalité" that can be rendered as "we must make life a dream and make a dream a reality."[43] Make reality dream, dream reality. This scientist is poetic and philosophical about his role as a scientist as well as about science itself. The natural philosopher thinks about natural philosophy but also about his own human nature and nature.

Marie interprets this passage in Pierre's diary. She praises the lucidity of his analysis for someone of twenty, evoking the environment for highest thought, providing "a true lesson" ("un véritable enseignement") making a way "for dreaming minds capable of opening new paths for humanity" ("aux esprits rêveurs capables d'ouvrir pour l'humanité des voies nouvelles").[44] Marie sees Pierre as an example, a model. For Marie, Pierre sought unity of thought but also had obligations in his profession and society and had tastes in culture – in art and literature. She also notes that Pierre, like his father, loved to read, including hard literary works, and his response to criticism for this was "I do not detest boring books" ("Je ne déteste pas les livres ennuyeux").[45] According to Marie, the "search for truth" ("la recherche de la vérité") that seduced him could be associated with "writing devoid of charm" ("une rédaction dépourvue d'agrément").[46] In the portrait of Pierre, a cultured man of the arts, Marie speaks of his love of music and painting. That commitment to culture also involved the practice of art: "Quelques fragments de poésies, transcrits de sa main, sont restés dans ses papiers" ("A few fragments of poetry, transcribed in his hand, remained among his papers").[47] He sought science and culture, so he could not focus fully on his scientific problems. Like his father, Pierre valued science and culture.

Marie Curie's verbal portrait of Pierre Curie is full of twists and turns and qualifications: it is an intricate representation. She says that all the concerns of Pierre that she has described were subordinate to his true task in his estimation. Moreover, Marie notes that Pierre felt incomplete when his scientific imagination was not fully active. She is frank when she introduces the relation between his feelings and words:

> Son inquiétude s'exprimait en paroles émouvantes, inspirées par la souffrance des périodes de depression passagères. «Que serai-je plus tard ? » a écrivait-il. « Bien rarement je suis tout à moi ; ordinairement une portion de mon être est endormie. Mon pauvre esprit, es-tu donc si faible que tu ne puisses réagir sur mon corps ? O mes pensées vous êtes donc bien peu de chose ! C'est en mon imagination que j'aurais le plus confiance pour me tirer de l'ornière, mais j'ai bien peur qu'elle ne soit morte.»[48]
>
> His disquiet was expressed in moving words, inspired by the suffering of passing periods of depression. 'What will I be later?' he wrote. 'Very rarely am I entirely my own; ordinarily, a portion of my being is dormant. My poor mind, are you so weak that you cannot affect [react on] my body? O my thoughts, you are then very little! It is in my imagination that I would have the most confidence to pull me out of the rut, but I am very afraid it is dead.'[49]

Marie is explicit about Pierre's disquiet and depression before she comes to his words, which, once more, she finds, so his self-representation can be part of the representation of his life. In these words, Pierre asks questions not unlike Hamlet, questions about being and mind. Pierre addresses or apostrophizes his thoughts. He appeals to imagination to pull him out of the rut yet expresses his fear that his imagination is dead.

Another of Marie Curie's turns in her portrait of Pierre Curie is a return to a focus on science. Despite the hesitations, doubts and loss, the young Pierre, she says, found his way and undertook fruitful research at an age when scientists are students. Beforehand, as I noted, Marie mentioned that at the Sorbonne in 1878, Paul Desains brought on Pierre Curie to assist in his laboratory and made him responsible for the laboratory work of the physics students. Here, Marie revisits that time in which Pierre collaborated with Desains on something novel, “the determination of heat wavelengths using a thermoelectric pile and a network of metal wires. This process, then completely new, has since been frequently used to study this subject” (“la détermination des longueurs d’onde calorifiques à l’aide d’une pile thermoélectrique et d’un réseau formé de fils métalliques”).[50] She gives details of the collaboration between Jacques Curie and Pierre Curie on crystals, their discovery of “*piezoelectricity* , which consists of an electrical polarization produced by the compression or expansion of crystals lacking a centre of symmetry” (“*piézoélectricité,* qui consiste en une polarisation électrique produite par la compression ou la dilatation des cristaux dépourvus de centre de symétrie”).[51] The *OED* defines the word, which came into English in 1883, “A phenomenon observed esp. in certain crystals whereby the application of mechanical stress causes a substance to become electrically polarized, and vice versa” and gives the etymology thus: “< **piezo-** *comb. form* + **electricity** *n.*, after **German** *Piezoelektricität* (W. G. Hankel 1881, in *Berichte der Königl. Sächsischen Ges. der Wissensch. zu Leipzig* vol. 33 52; now *Piezoelektrizität*). Compare **French** *piézoélectricité* (1890), **Italian** *piezoelettricità* (1900).”[52] The *OED* traces the word to German and W. G. Hankel in German, but he coined the term to describe what the Curie brothers had discovered, as Antonio Arnau and David Soares explain:

> The word *Piezoelectricity* comes from Greek and means ‘electricity by pressure’ (*Piezo* means pressure in Greek). This name was proposed by Hankel [1] in 1881 to name the phenomenon discovered a year before by the Pierre and Jacques Curie brothers [2]. They observed that positive and negative charges appeared on several parts of the crystal surfaces when comprising the crystal in different directions, previously analysed according to its symmetry.[53]

Marie’s analysis attributes this discovery to the brothers reflecting “on the symmetry of crystalline matter” (“sur la symétrie de la matière crystalline”), which enabled Jacques and Pierre “to foresee the possibility of this polarization” (“de prévoir la possibilité de cette polarization”).[54] The brothers Curie showed skill in experiments and studied the new phenomenon while “establishing the symmetry conditions necessary for its production in crystals, and describing its remarkably simple quantitative laws, as well as its absolute magnitude for certain crystals” (“ont établi les conditions de symétrie nécessaires à sa production dans les cristaux, ont donné ses lois quantitatives remarquablement simples, ainsi que sa grandeur absolue pour certains cristaux").[55] Marie also notes that others outside France, such as William Rontgen [Roentgen], August Kundt, Wilhelm Voigt, and Karl Victor Eduard Riecke, pursued research in the area that Jacques and Pierre opened up. She also observes that the second and more difficult phase of this work from the point of view of experiment concerns deformation by piezoelectric crystals "when subjected to the action of an electric field" ("quand ils sont soumis à l'action d'un champ électrique"), which Lippmann predicted and Jacques and Pierre Curie demonstrated.[56] For Marie Curie, the smallness of the deformations necessary for observation made this research difficult. The difficulty lay in the smallness of the deformations that needed to be observed.

Being an accomplished scientist, Marie assesses the contribution the brothers Curie made. She says that from this theoretical and experimental research, they deduced the *piezoelectric quartz*, a new device to "measure, in absolute value, small quantities of electricity and low-intensity electric currents" ("mesurer, en valeur absolue, de faibles quantités d'électricité"), a device that "later provided great service in research on radioactivity" ("a, plus tard, rendu de grands services dans les recherches sur la radioactivité").[57] Something Marie Curie says in a footnote is that there has been a recent application for the piezoelectric properties of quartz. Paul Langevin, Pierre's doctoral student, employed it to produce ultrasound or high-frequency elastic waves "emitted in water, for the purpose of detecting underwater obstacles. This same method can be used more generally to explore the depths of the sea" ("émises dans l'eau, dans le but de la detection d'obstacles sons-marins. Cette même méthode peut server d'une manière plus générale à explorer les profondeurs de la mer"); and she makes a general observation – to end this note – not simply for her times but beyond: "We thus see, once again, how pure speculation can lead to a discovery that can later be used in unforeseen directions" ("On voit ainsi, une fois de plus, comment la speculation pure peut conduirca unc découverte utilisable plus tard dans des directions imprévues").[58] Marie also stresses the friendship of the brothers, their collaboration and "their passion for science" ("leur passion pour la science").[59] She characterizes the brothers: "la vivacité et l'énergie de Jacques étaient un secours précieux pour Pierre, plus aisément absorbé dans ses pensées" ("the vivacity and the energy of Jacques were a precious help to Pierre, who was more easily absorbed in his thoughts").[60] That Jacques was full of life was a gift to the contemplative Pierre. Marie recounts that the brothers collaborated for a few years, Jacques going to Montpellier in 1883 and Pierre remaining in Paris to head up the School of Industrial Physics and Chemistry of the city of Paris: Jacques and Pierre Curie's work on crystals earned them the Gaston Planté Prize of the Academy of Sciences in 1895.[61]

The Curie family, Jacques, Pierre, Marie and beyond, helps to make significant scientific discoveries. In the next chapters, I discuss the last day of Pierre's life and the life of Marie. The emphasis moves from representation to self-representation.

Notes

1 Marie Curie, *Pierre Curie*, translated by Charlotte Kellogg and Vernon L. Kellogg, With an Introduction by Mrs. William Brown Meloney and Autobiographical Notes by Marie Curie (New York: The Macmillan Company, 1923); Marie Curie, *Pierre Curie* (Les grands hommes de France, collection dirigée par Paul Gautier) (Paris: Payot, 1924). The electronic versions of the English text I am using have either no pagination or contain pages without pagination. I shall try to find and use page numbers when possible.
2 Curie, *Pierre Curie* (Les grands hommes de France, ...), [no pagination]. The document, including this beginning of the concluding paragraph, is available at the website of Le Musée Curie (Institut Curie) at https://musee.curie.fr/.uploads/2018-01/nobel-1903-c6fa2633.pdf
3 Pierre Curie quoted in Marie Curie, *Pierre Curie*, translated by Kellogg and Kellogg (1923), [3].
4 Curie, "Preface," *Pierre Curie*, 7.
5 Marie Curie, *Pierre Curie* (Les grands hommes...), [no pagination].
6 Marie Curie, "Preface," *Pierre Curie*, 7.
7 Marie Curie, "Preface," *Pierre Curie*, 7.
8 Curie, "Preface," *Pierre Curie*, 7.
9 Curie, *Pierre Curie* (Les grands hommes...), [no pagination].
10 Curie, "Preface," *Pierre Curie*, 7.
11 Curie, "Preface," *Pierre Curie*, 7.
12 Curie, *Pierre Curie* (Les grands hommes...), [no pagination].
13 Curie, "Preface," *Pierre Curie*, 8.
14 Curie, *Pierre Curie* (Les grands hommes...), [no pagination].
15 Curie, *Pierre Curie* (Les grands hommes...), [no pagination].
16 Curie, *Pierre Curie*, 28.
17 *Oxford English Dictionary*, "preparator (*n.*)," July 2023, https://doi.org/10.1093/OED/1448514387.
18 Curie, *Pierre Curie* (Les grands ...), 10.
19 Curie, *Pierre Curie* (Les grands ...), 10.
20 Curie, *Pierre Curie* (Les grands ...), 13; my literal translation; for the English version of the book, see Curie, *Pierre Curie*. 33.
21 Curie, *Pierre Curie* (Les grands ...), 14; my translation; see Curie, *Pierre Curie*. 34.
22 Curie, *Pierre Curie* (Les grands ...), 14.
23 Curie, *Pierre Curie* (Les grands ...), 14; see Curie, *Pierre Curie*, 34–35.
24 Curie, *Pierre Curie* (Les grands ...), 15.
25 My translation; see also Curie, *Pierre Curie*, 35.
26 Curie, *Pierre Curie* (Les grands ...), 15; Curie, *Pierre Curie*, 35–36.
27 Curie, *Pierre Curie* (Les grands ...), 16; Curie, *Pierre Curie*, 36.

28 Curie, *Pierre Curie* (Les grands ...), 16; Curie, *Pierre Curie*, 37.
29 Curie, *Pierre Curie* (Les grands ...), 17; my translation; see Curie, *Pierre Curie*, 37.
30 Curie, *Pierre Curie* (Les grands ...), 17; my translation; see Curie, *Pierre Curie*, 37.
31 Curie, *Pierre Curie* (Les grands ...), 17. I have corrected the typographical error "Pusieurs" in the original.
The French source is inconsistent in sometimes having and sometimes not a space before the following punctuation marks : ; ? ! I have tried to reproduce those marks as they occur in the passage quoted in the chapters on the Curies.
32 My translation; see Curie, *Pierre Curie*, 37–38.
33 Curie, *Pierre Curie* (Les grands ...), 17; my translation; see Curie, *Pierre Curie*, 38.
34 My translation; Curie, *Pierre Curie* (Les grands ...), 18; see Curie, *Pierre Curie*, 40; see 39.
35 Curie, *Pierre Curie* (Les grands ...), 19.
36 My translation; see Curie, *Pierre Curie*, 40.
37 Curie, *Pierre Curie* (Les grands ...), 20; my translation; see Curie, *Pierre Curie*, 41.
38 Curie, *Pierre Curie* (Les grands ...), 20; my translation; see Curie, *Pierre Curie*, 42.
39 My translation; Curie, *Pierre Curie* (Les grands ...), 21, see 20; see also Curie, *Pierre Curie*, 42.
40 Curie, *Pierre Curie* (Les grands ...), 21; my translation; Curie, *Pierre Curie*, 41.
41 Curie, *Pierre Curie* (Les grands ...), 21–22; my translation; see Curie, *Pierre Curie*, 43.
42 Curie, *Pierre Curie* (Les grands ...), 22; my translation; see Curie, *Pierre Curie*, 44.
43 Curie, *Pierre Curie* (Les grands ...), 22; my translation; see Curie, *Pierre Curie*, 44.
44 My translation; Curie, *Pierre Curie* (Les grands ...), 22; see Curie, *Pierre Curie*, 44.
45 My translation; Curie, *Pierre Curie* (Les grands ...), 23; see Curie, *Pierre Curie*, 44.
46 My translation; Curie, *Pierre Curie* (Les grands ...), 23; see Curie, *Pierre Curie*, 44–45.
47 Curie, *Pierre Curie* (Les grands ...), 23; my translation; see Curie, *Pierre Curie*, 45.
48 Curie, *Pierre Curie* (Les grands ...), 23.
49 My translation; see Curie, *Pierre Curie*, 45.
50 My translation; Curie, *Pierre Curie* (Les grands ...), 23–24; see Curie, *Pierre Curie*, 46.
51 My translation; Curie, *Pierre Curie* (Les grands ...), 23–24; see Curie, *Pierre Curie*, 46.
52 *Oxford English Dictionary*, "piezoelectricity (*n.*)," March 2025.
53 Antonio Arnau and David Soares, "Fundamentals of Piezoelectricity," in *Piezoelectric Transducers and Applications*, edited by Antonio Arnau Vives (Berlin and Heidelberg: Springer, 2009), 2, see 1. See also Pierre and Jacques Curie, "Développement, par pression, de l'électricité polaire dans les cristaux hémièdres à faces inclinées," *Comptes Rendus* 91 (1880): 294–295; Wilhelm Gottlieb Hankel, "Uber die aktinound piezoelektrischen eigenschaften des bergkrystalles und ihre beziehung zu den thermoelektrischen," *Abh. Söchs* 12 (1881): 457.
54 My translation; Curie, *Pierre Curie* (Les grands ...), 24; see Curie, *Pierre Curie*, 46.
55 My translation; Curie, *Pierre Curie* (Les grands ...), 24; see Curie, *Pierre Curie*, 46.
56 My translation; Curie, *Pierre Curie* (Les grands ...), 24; see Curie, *Pierre Curie*, 47.
57 My translation; Curie, *Pierre Curie* (Les grands ...), 25; see Curie, *Pierre Curie*, 47.
58 My translation; Curie, *Pierre Curie* (Les grands ...), 25; see Curie, *Pierre Curie*, 47–48.
59 My translation; Curie, *Pierre Curie* (Les grands ...), 25; see Curie, *Pierre Curie*, 48.

60 Curie, *Pierre Curie* (Les grands ...), 26; my translation; see Curie, *Pierre Curie*, 48.
61 Curie, *Pierre Curie* (Les grands ...), 26; see Curie, *Pierre Curie*, 48–49. The Academy of Sciences in Paris in a publication, pages 1059–1061, records in some detail the research leading to this prize for the Curie brothers (see https://www.academie-sciences.fr/pdf/dossiers/Curie/Curie_pdf/Curie_Plante_CRT121.pdf).

7 Marie Curie

The Life of Pierre Curie – Scientific Contributions and Marriage

Marie turns from the youth of Pierre to stress his strides in science and his marriage and family. Struggles and challenges continue. The life of this scientist is not easy, not simply in terms of intellectual effort but also because of the pressures that life brings.

Marie's third chapter is about the early research at the School of Physics on symmetry and magnetism. Her life of Pierre Curie becomes, increasingly, about the scientific work in which Pierre is engaged. Marie speaks about the old buildings in which he worked for twenty-two years, and she quotes his own words at the end of a lecture at the Sorbonne in the last years of his life. They conducted their research at the School of Physics and Chemistry of the City of Paris. Pierre stresses the importance of environment in science. He expresses his gratitude to Schützenberger, the first director of the School, for supporting him and Madame Curie, who was able "to come and work near me, and this permission, at the time it was given, was an unusual innovation. Schützenberger allowed us all great freedom, and his influence was felt above all through his infectious passion for science" ("de venir travailler près de moi, et cette autorisation, à l'époque à laquelle elle a été donnée, était une innovation peu ordinaire. Schützenberger nous laissait à tous une grande liberté, et son action se faisait sentir surtout par sa passion communicative pour la science").[1] Pierre praises the environment, the professors, the students and the former students who are collaborators and friends and expresses his gratitude.

Marie introduced words from Pierre's speech with her own interpretation, and now she continues in that vein. She humanizes him and describes him as a young teacher barely older than his students. She provides an illustrative anecdote, a small narrative within a large narrative, in French *un petit récit dans un grand récit*. In this account, Marie speaks of Pierre smiling in recalling from this era when Pierre and his students found the door shut and they had to descend from the first-floor single file through a pipe by the window. The context of this recollection was a dinner of an association of former students in 1903, and in part of a larger framework in which Marie shows a bond between the students and Pierre Curie. Marie had told of how

DOI: 10.4324/9781003651871-8

much the students had loved Pierre Curie, the new director of the laboratory, whose simple manners and youth made him more a comrade than a master ("Il était, au début, dans ses nouvelles fonctions, à peine plus âgé que ses élèves qui l'aimaient en raison de l'extrême simplicité de ses manières, celles d'un camarade autant que d'un maître").[2] She depicts the students as full of emotion recalling their sessions with Pierre at the backboard and the enthusiasm that came with their scientific discussions. Personal and familial bonds of feeling are important in science and the pursuit of scientific matters. Marie focuses on Pierre's character and elaborates on similar points, a kind of characterization through amplification. She says that Pierre's reserve and shyness made it hard for him to make friends, but that those who worked with him loved Pierre owing to his kindness. Some of her observations are not simply about closeness but also, consciously or not, hierarchical:

> Tel a été pendant toute sa vie le cas de ses subordonnés. A l'Ecole, le garçon de laboratoire de son service, qu'il avait eu l'occasion d'aider dans une circonstance difficile de sa vie, avait pour lui une grande reconnaissance et un véritable culte (This was the case with his subordinates throughout his life. At the School, the laboratory assistant in his department, whom he had the opportunity to help in a difficult circumstance in his life, had for him a great gratitude and a veritable cult).[3]

Greatness in gratitude might be more desirable than a cult, although Marie may be trying to represent Pierre as a quiet hero and a shy but inspiring leader. Looking at this representation in the most positive way, one might say that the portrait of a shy and kind Pierre is also one of mutual kindness with those close to him in family and at work.

Jacques Curie comes back into the narrative. Marie notes that even though the brothers were separated, Pierre kept the bond of friendship and trust with Jacques, who collaborated with Pierre during holidays, coming to Paris, but Pierre sometimes went to meet Jacques to help him with geological mapping work in Auvergne and accompanied him on walks each day needed to draw the map. Now Marie injects a personal note that brings her into the story of the brothers.

She includes memories of one of these hikes, excerpting from a letter Pierre sent Jacques shortly before the wedding with Marie. The passages from the letter begin with Pierre's expression of how happy he was to spend time with his brother then saying how they were far from worries and did not know where they would sleep the following day. Pierre writes:

> Par moments, il me semblait être revenue à l'époque où nous vivions toujours ensemble. Nous en étions alors arrivés à avoir, sur toutes choses,

les mêmes opinions à ce point que, pensant de même, il ne nous était plus nécessaire de parler pour nous comprendre. Cela était d'autant plus étonnant que nous avons des caractères entièrement différents (At times, it seemed to me that I had returned to the time when we always lived together. We had then come to have, on all things, the same opinions to such an extent that, thinking alike, it was no longer necessary for us to speak to understand each other. This was all the more astonishing since we have entirely different personalities).[4]

Pierre stresses, and Marie through him, a return to the former times when the brothers lived nearby and worked together and the renewed bond between brothers so different in personality, sharing thoughts unspoken. Marie, an eminent scientist, takes time out to write a biographical account of her dead husband and then is asked to provide autobiographical notes. Personal and family reputation and honour seem to be at stake. Reputation is important for scientists as for others.

At first, Marie says, Pierre Curie's appointment to the School of Physics and Chemistry initially delayed his experimental research because this establishment was created anew and nothing yet existed there. Thus, Pierre organized the experimental department with "his characteristic spirit of precision and innovation" ("l'esprit de précision et de nouveauté qui le caractérisait").[5] As head of research, Pierre worked hard and oversaw the many experiments of the students and used this interruption of his experimental research to finish his scientific education, especially his mathematical knowledge while engaging in theoretical reflections on, as Marie notes, "the connections between crystallography and physics" ("les liaisons qui existent entre la cristallographie et la physique").[6] Marie then discusses Pierre's work in 1884 and 1885, including the publication of "a very important theoretical work on the formation of crystals and on the capillary constants of the different faces" ("il publia un travail théorique très important sur la formation des cristaux et sur les constantes capillaires des différentes faces").[7] All this, Marie says, shows Pierre's preoccupation with crystal physics; his theoretical and experimental research is grouped around the principle of symmetry, which was published in its final form between 1893 and 1895, and she records it in his words. She then compresses his statement into a helpful expression: "L'importance capitale de cet énoncé, très parfait dans sa simplicité, consiste en ce que les éléments de symétrie qu'il fait intervenir sont relatifs *à tous les phénomènes sans exception*" ("The crucial importance of this statement, very perfect in its simplicity, lies in the fact that the elements of symmetry it involves are relative *to all phenomena without exception*").[8] Marie discusses physics, including Pierre's contribution, and does so in detail and places his work in context. After her discussion, Marie stresses the accomplishments of Pierre once more:

> 'On peut se rendre compte, par cet exposé sommaire, de la haute portée philosophique de ces notions de symétrie, qui interviennent dans tout phénomène naturel, et dont le sens profond a été dégagé d'une manière si efficace par la pensée claire de Pierre Curie' ('This brief statement gives an idea of the profound philosophical significance of these notions of symmetry, which are present in every natural phenomenon, and whose profound meaning has been so effectively brought out by the clear thinking of Pierre Curie').[9]

Marie speaks about how significant philosophically his ideas of symmetry are. Her emphasis is discrete but also cumulative and she builds his portrait in words through repetition with variation.

Marie Curie recalls and quotes Louis Pasteur's view of "The Universe, said he, is an asymmetrical whole" ("L'Univers, disait-il, est un ensemble dissymétrique").[10] Pierre studied symmetry and asymmetry. Marie wishes to show the challenges that Pierre overcame to make his experimental discoveries. She notes that he did not have his own laboratory, rooms or research funding until, after several years at the School, he obtained, owing to the support of Schützenberger, a small annual grant. According to Marie Curie, Pierre conducted some of his experiments in the student rooms when not in use or more often "in a narrow passage between a staircase and a room for experiments" ("dans un passage exigu compris entre un escalier et une salle de manipulations") where he did his long-time work on magnetism.[11] Pierre took something positive from these unfavourable circumstances for the scientist: they brought him closer to his students, who sometimes participated in his science. Marie notes that his return to experimental research involved the study "of the 'Aperiodic Precision Balance with Direct Reading of the Last Weights' (1889, 1890, 1891)" ("une étude approfondie de la 'Balance de précision apériodique à lecture directe des derniers poids' (1889, 1890, 1891)" and after explaining the matter, she says that this work "on this subject is far from empirical but includes a theoretical study of damped movements and the construction of numerous curves established with the help of some of his students" ("à ce sujet est loin d'être empirique if comporte une étude théorique des mouvements amortis et la construction de nombreuses courbes établies avec l'aide de quelques-uns de ses élèves").[12] About 1891, Marie says, Pierre began research on the magnetic properties of bodies at various temperatures. She explains that in 1895 this work was presented as a doctoral thesis to the

Faculty of Sciences in Paris and quotes his explanation of it, part of which is as follows:

> My experiments did not produce any similarity between the properties of diamagnetic bodies and those of paramagnetic bodies, and the *results support theories that attribute magnetism and diamagnetism to causes of a different nature*. On the contrary, the properties of ferromagnetic bodies and weakly magnetic bodies are intimately related (Mes expériences n'ont amené aucun rapprochement entre les propriétés des corps diamagnétiques et celle des corps paramagnétiques, et les *résultats sont favorables aux théories qui attribuent le magnétisme et le diamagnétisme à des causes de nature différente*. Au contraire, les propriétés des corps ferromagnétiques et des corps faiblement magnétiques sont reliées intimement).[13]

Marie stresses the experimental difficulties and notes that Pierre Curie understood that the results were significant from a theoretical point of view, observing that

> *Curie's law*, according to which the magnetization coefficient of weakly magnetic bodies varies inversely with the absolute temperature, is a remarkably simple law, entirely comparable to Gay-Lussac's law relating to the variation of the density of an ideal gas with temperature (*La loi de Curie*, d'après laquelle le coefficient d'aimantation des corps faible-ment magnétiques varie en raison inverse de la temperature absolue, est une loi remarquablement simple, entièrement comparable à la loi de Gay-Lussac relative à la variation de la densité d'un gaz parfait avec la temperature).[14]

Marie explains how Paul Langevin and Pierre Weiss helped to show the importance of Pierre's work.

Free inquiry is one of the themes that Marie Curie stresses when portraying Pierre Curie. He searched for new phenomena, but without results or publications, "having the habit of engaging in the pursuit of phenomena, often without much hope of success, solely out of a love of the unforeseen, and without holding on in any way to an eventual publication" ("ayant l'habitude de s'engager ainsi dans la poursuite de phénomènes, souvent sans grand espoir de succès, par unique amour de l'imprévu, et sans tenir en aucune manière à une publication éventuelle").[15] There is an otherworldly or unambitious aspect to Marie's representation of Pierre. She speaks of an "entirely disinterested passion for scientific research" ("Cette passion entièrement désintéressée pour la recherche scientifique"), so he did not gather his work on magnetism for his doctoral thesis until he was thirty-five.[16] An entirely disinterested passion seems to be an oxymoron but makes sense if it is a passion for truth or fact or scientific discovery, although passion and reason are opposites that need to coexist in some balance. Marie seems to be invoking passion as a driving and pure motivation for Pierre's scientific work.

Personal memory is also part of Marie's biographical work: "Je conserve un souvenir très vivant de cette soutenance de thèse, à laquelle il m'invita, en raison de l'amitié qui nous liait déjà à cette époque" ("I conserve a very vivid memory of this thesis defence, to which he invited me, because of the friendship that already bound us at that time").[17] Marie tells of the jury, Pierre's friends and father who attended, of the simple, clear presentation, of the esteem the professors showed and is not afraid of showing her own passion for science: "La petite salle abrita, ce jour-la, la haute pensée humaine, et de ce sentiment j'étais toute pénétrée" ("The small room housed, that day, the highest human thought, and with this feeling I was completely penetrated").[18] Friends, a proud father and professors respecting Pierre, who is once more made clear and simple in his thinking – straightforward; all this underscores her feeling of this conversation about the heights of human science.

Here, Marie adds a summary. She stresses the progress in Pierre's life between 1883 and 1895 when as a young physicist he was a research leader. He organized a new teaching department, published theoretical papers, did experimental research and constructed a new apparatus in inadequate conditions and without adequate funding. The esteem for Pierre was growing in France and beyond. Lord Kelvin was one of his admirers, and Marie tells an anecdote in which in a meeting in person Kelvin comes to admire the accuracy of Pierre's work. Marie also stresses the modesty of Pierre's post and one of the reasons she gives for this position: "His independence of character" ("L'indépendance de son caractère").[19] He also thought, according to Marie, that seeking honours was harmful. For instance, when Schützenberger wanted to nominate him for the Palmes Académiques, he refused this distinction. Marie focuses on advancement, work conditions, honours all in relation to Pierre's character. She says that Éleuthère Mascart, a professor of physics at the Collège de France, esteemed Pierre and, knowing that Lord Kelvin also did, suggested to Schützenberger that he request a new chair of physics at the School.

In Chapter 4, Marie discusses marriage, family life, personality and character. She begins with an account of meeting Pierre Curie in the spring of 1894 in Paris, where she had been studying at the Sorbonne for three years. She says that she had passed the examinations for a physics degree and was preparing for the degree in mathematics. She tells of a Polish physicist who invited Pierre and her, Marie Sklodowska (which she explains in a brief autobiographical footnote), one day to spend the evening with his wife and him. Marie describes her first sight of Pierre Curie. At first sight, she could see his clear gaze with "a slight air of abandonment" ("une légère apparence d'abandon"), his tallness, unpretentious and thoughtful speech, simplicity, smile.[20] Some of these qualities, including simplicity, she has amplified in representing Pierre before he met her. They soon had a friendly conversation about scientific matters she posed and then social and humanitarian questions in which they were both interested. She expresses her surprise:

> Il y avait entre sa conception des choses et la mienne, malgré la différence de nos pays d'origine, une parenté surprenante, attribuable, sans doute, en partie, à une certaine analogie dans l'atmosphère morale, au milieu de laquelle chacun de nous avait grandi dans sa famille (There was a surprising similarity between his conception of things and mine, despite the difference in our countries of origin, attributable, without a doubt, in part, to a certain analogy in the moral atmosphere in which each of us had grown up in his [or her] families).[21]

In this first meeting, Marie is astonished at the similarity between them, and she looks to families and their moral views for an explanation.

Marie sees their connection developing through science at work, but then Pierre, who showed sympathy for her work life, visited her in her modest accommodation on the sixth floor, where, despite her lack of resources, she was happy, "having realized at only twenty-five the ardent desire I had long had to pursue advanced scientific studies" ("ayant réalisé à vingt-cinq ans seulement l'ardent désir que j'avais eu depuis longtemps de faire des études de sciences approfondies").[22] Marie is now part of the story, not just for her but also for Pierre. She represents this pivotal moment: "Bientôt il prit l'habitude de me parler de son rêve de vie consacrée entièrement à la recherche scientifique, et il me demanda de partager cette existence" ("He soon had the habit of talking to me about his dream of a life consecrated entirely to scientific research, and he asked me to share this existence").[23] The dream of a shared consecration of life to science is central to the biography as well as to this key point in the narrative. Marie also notes that this decision was not easy because she was separated from country and family, and would be difficult, given her social activity: "Ayant grandi dans une atmosphère de patriotisme entretenue par l'oppression exercée sur la Pologne, je voulais, comme tant d'autres jeunes gens de mon pays, contribuer par mes efforts à la conservation de l'esprit national" ("Having grown up in an atmosphere of patriotism maintained by the oppression exercised in Poland, I wanted, like so many other young people in my country, to contribute through my efforts to the preservation of the national spirit").[24] Family, culture, politics and society at home in Poland mean a great deal to Marie, so this decision to be with Pierre in Paris into the future is not simple. Science, as much as she is devoted to it, does not occur in a vacuum.

Here, and elsewhere in the biography, Marie represents how Pierre speaks and writes. She tells how at the start of the vacation, she left Paris for Poland to visit her father. She attributes to the correspondence between Pierre and her something that "helped to strengthen the bond of affection that was beginning to be established between us" ("a contribué à ressèrrer entre nous le lien d'affection qui commençait à s'établir").[25] Moreover, Marie praises the letters Pierre wrote her during the summer of 1894, and she summarizes their nature and comments on the nature of the writing before she introduces the reason for quoting from them:

> Pierre Curie m'écrivit durant l'été 1894 des lettres que je crois admirables dans leur ensemble. Aucune d'elles n'est très longue, car il avait l'habitude de la rédaction concise, mais toutes ont été écrites par lui dans le souci évident de se faire connaître à celle qu'il désirait pour compagne, tel qu'il était, avec une sincérité objective. La qualité même de la rédaction m'a toujours paru exceptionnelle ; nul n'était capable comme lui de décrire en peu de lignes un état d'esprit où une situation, de manière à en évoquer une image saisissante de vérité par des moyens très simples.
>
> Pierre Curie wrote me letters during the summer of 1894 that I believe to be admirable overall. None of them is very long, as he had the habit of concise writing, but all were written by him with the obvious concern of making himself known to the woman he desired as a companion, as he was, with objective sincerity. The very quality of his writing has always seemed exceptional to me; no one was as capable as he was of describing in a few lines a state of mind or a situation, in such a way as to evoke a strikingly image of truth by very simple means.[26]

She praises Pierre's writing, its concision, objectivity, sincerity, image, truth, simplicity. For Marie, as for Pierre, writing and communication are key matters. Making her own method explicit, Marie selects and orders, saying that she has quoted from Pierre's letters: "Quelques fragments de ses lettres ont déjà été cités au cours de ce récit, et d'autres le seront par la suite. Il convient d'en reproduire ici quelques passages qui expriment comment il envisageait l'éventualité de son marriage" ("Some fragments of his letters have already been quoted in this story, and others will be quoted later. It is appropriate to reproduce here a few passages that express how he envisaged the possibility of his marriage").[27] She sees herself as telling "this story" ("ce récit") and decides to share these private words between Pierre and her publicly, with the reader in a book. Marie has a keen eye and selects a passage that is a key to the relation between science and society:

> Nous nous sommes promis (n'est-il pas vrai ?) d'avoir l'un pour l'autre au moins une grande amitié. Pourvu que vous ne changiez pas d'avis ! Car il n'y a pas de promesses qui tiennent ; ce sont des choses qui ne se commandent pas. Ce serait cependant une belle chose à laquelle je n'ose croire que de passer la vie l'un près de l'autre, hypnotisés dans notre rêve : *votre rêve* patriotique, *notre rêve* humanitaire et *notre rêve* scientifique. De tous ces rêves-là, le dernier seul, est, je crois, légitime. Je veux dire par là que nous sommes impuissants pour changer l'état sociale et s'il n'en était pas ainsi, nous ne saurions que faire, et en agissant dans un sens quelconque nous ne serions jamais sûrs de ne pas faire plus de mal que de bien, en retardant quelque évolution inévitable. Au point de vue scientifique, au contraire, nous pouvons prétendre faire quelque chose ; le terrain est ici plus solide et toute découverte, si petite qu'elle soit, reste acquise.
>
> We promised each other (isn't that true?) to have at least for each other a great friendship. Provided you do not change your mind! For there are no promises that keep; they are things that cannot be ordered. It would, however, be a beautiful thing, one I dare not believe, to spend our lives close to each other, hypnotized by our dreams: *your* patriotic *dream*, *our* humanitarian *dream*, and *our* scientific *dream*. Of all these dreams, only the last is, I believe, legitimate. I wish to say by this that we are powerless to change the social state, and if it were not so, we would not know what to do, and by acting in any direction we would never be sure of not doing more harm than good by retarding some inevitable evolution. From a scientific point of view, on the contrary, we can claim to be doing something; the ground is more solid here, and any discovery, however small, remains a given.[28]

Pierre speaks about dreams but sees that the only legitimate one is science. Patriotism and humanitarian dreams reveal the powerlessness of Pierre and Marie, their unsureness about doing more harm than good as opposed to science doing something good and allowing for a discovery no matter how modest. Marie quotes more from the letter in which Pierre advises her to return in October to Paris and says his view is not selfish because he thinks she will work better in Paris. She interprets the excerpt from this letter, as she did in introducing it:

> On peut comprendre, d'après cette lettre, qu'il n'y avait pour Pierre Curie qu'une seule manière d'envisager son avenir. Il avait voué sa vie à son rêve scientifique ; il lui fallait une compagne qui pût vivre le même rêve avec lui (We [one] can understand, according to this letter, that there was only one way for Pierre Curie to envision his future. He had devoted his life to his scientific dream; he needed a companion who could live the same dream with him).[29]

This repetition stresses the dream of science and how Pierre said he did not marry because he needed science, and it seems that Marie's love of science allowed them both to pursue that dream together.

There are many intriguing details in Marie's biography of Pierre, and, for the sake of space, I have to be selective. She quotes Louis Pasteur – words, she says, that are well known at the time – as a gloss on Pierre's commitment to science: "Je crois invinciblement que la science et la paix triompheront de l'ignorance et de la guerre" ("I believe invincibly that science and peace will triumph over ignorance and war").[30] Marie sees Pierre's commitment to science for humanity. This is a crux for Pasteur and the Curies, as well as for others like Darwin and Russell, and is important for my book, which looks at science and culture, the human and literary elements in what was once called natural philosophy and in society.

Pierre, Marie says, did not take an active role in politics, was attached to education and democratic and socialist ideas but not to any party doctrine while fulfilling, as his father did, his obligations as an elector (voter): "Dans la vie publique comme dans la vie privée, il ne croyait pas à l'emploi de la violence" ("In both public and private life, he did not believe in the use of violence").[31] Pierre is a gentleman. Although Marie's portrait of Pierre has him live the dream of science, her representation also includes his realistic views of society. She quotes his words to her, some of which I quote here:

> Je crois encore que la justice n'est pas de ce monde, et que le systèmele plus fort, ou plutôt le plus économique sera celui qui prévaudra. Un homme est exténué par le travail et vit quand même misérable c'est là une chose révoltante, mais ce n'est pas pour cela qu'elle cessera elle disparaîtra probablement parce que l'homme est une sorte de machine, et qu'il y a avantage au point de vue économique à faire fonctionner une machine quelconque dans son régime normal, sans la forcer.
>
> I still believe that justice is not of this world, and that the strongest, or rather the most economical, system will be the one that prevails.

> A man is exhausted by work and still lives in misery; this is a revolting thing, but it is not for that it will cease; it will probably disappear because man is a kind of machine, and there is an advantage from the economic point of view in making any machine function in its normal regime, without forcing it.[32]

Pierre's view of economics is that in an economy, the human is a machine, and profit dictates that if one forces the machine beyond its usual range, that will limit profit. Marie also quotes from his writing that involves introspection in which he declares that we are slaves to passion and to the prejudices of those we love: "nous devons aussi gagner notre vie, et par cela devenir un rouage de machine" ("we must also earn our living, and in doing so become a cog in the machine").[33] How to be human in a society with an economy that makes people into machines? That is a question that haunts us still. To Marie, Pierre wrote that the most pain comes from making concessions we must make to social prejudices, and what we concede depends on how strongly we feel. For Pierre, if we concede too little, "we are crushed" ("on est écrasé"), but if we concede too much, "we are vile and become disgusted with ourselves" ("on est vil et l'on prend le dégoût de soi-même").[34] Pierre adds that he is far from his principles ten years before when he believed in excess and conceding nothing to the milieu, exaggerating faults as if they were qualities. Marie says that these were Pierre's thoughts, of someone without fortune to a student without fortune. She returned from her vacation and says that they both realized that there was no better companion for life, and they set their marriage for 25 July 1895. Marie describes their first and modest accommodations near the School of Physics, furnished with what their parents gave them, and how she had to do the housework. She also tells of Pierre's modest salary and how she was studying for an examination for young women to become teachers, and she passed that in 1896. Keeping to her theme, Marie says that they organized their lives according to scientific work, spending days in the laboratory, where Schützenberger allowed her to work with Pierre.

She explains that Pierre focused on crystal growth, but interrupted that work to concentrate on radioactivity, and regretted that he never returned to it, while Marie studied the magnetization of hardened steels. Regret and unfinished and unpublished work are part of Marie's portrait of Pierre, who died in his prime from an accident. Marie speaks of his teaching and the care he took in crystallography and electricity. She then mentions a theoretical course in electricity for future engineers, which Marie praises for Pierre's effort and clarity and precision in his presentation of phenomena, theories and ideas. Moreover, she says that he was concerned with precision and clarity in the presentation. She states that he wanted to publish a treatise that summarized this teaching but was unable to do so because of his many concerns. Their life together is now a key theme in Marie's account:

> Nous vivions très unis, nous intéressanten commun à toutes choses travail théorique, expériences de laboratoire, préparation de cours ou d'examens. Pendant onze années de vie commune, nous ne nous sommes presque pas séparés, à tel point qu'il n'existe que peu de lignes de correspondance entre nous de cette époque (We lived very close, sharing common interests in everything: theoretical work, laboratory experiments, and preparing for lectures or examinations. During eleven years of living together, we were almost inseparable, so much so that there are only a few lines of correspondence between us from that time).[35]

Scientific work was central.

Even when Marie describes their days off or vacations, she stresses Pierre's preoccupation with work, but she mentions the walks he and she took together, just as he had with his brother, being in the beauty of nature away from scientific questions.

These days, she said, they liked to recall, and she becomes poetic: "Un souvenir radieux nous est resté d'une journée de soleil, où, après une montée longue et pénible, nous traversions la prairie verte et fraîche de l'Aubrac, dans l'air pur des hauts plateau" ("A radiant memory remains of a sunny day, where, after a long and arduous climb, we crossed the fresh, green meadows of Aubrac, in the pure air of the high plateaus").[36] Marie herself can write beautifully, and her descriptions of nature and their experience on these walks, of which I have given a fragment, are memorable. After these beautiful descriptions, Marie shifts point of view and focuses on Pierre: "à la lisière de la forêt de Fontainebleau, les bords du Loing, chargés de renoncules d'eau, étaient pour Pierre Curie un objet de ravissement" ("on the edge of the Forest of Fontainebleau, the banks of the Loing, laden with water buttercups, were an object of delight for Pierre Curie").[37] Plants brought delight to Pierre. Marie then mentions their child and vacations at the same place.

The question of family is never far from the mind of Marie. Earlier, she had mentioned that Pierre's family and hers were at the wedding. Marie talks about the good relations she has with Pierre's parents, Jacques Curie and family, that a few days after the birth of her daughter, Irène, in September 1897, Pierre's mother died. At the house of his parents in Paris, Pierre lived until the end of his life. Marie speaks about how hard it was for Pierre and her to work with the birth of their daughter, but Pierre's father, a doctor and now a widower, looked after his granddaughter. Pierre and Marie worked on radioactivity: Marie says Pierre was social with his friends from childhood and those interested in science. Among the friends are E. Gouy, Ch.-Ed. Guillaume, André Debierne, Georges Sagnac, Paul Langevin, Jean Perrin and Georges Urbain, and she says that there were gatherings at the Curie house on Boulevard Kellermann: "Nous poursuivions alors une causerie sur les expériences récentes ou futures, sur les idées et théories nouvelles, et nous ne nous lassions pas de jouir du merveilleux développement de la physique modern" ("We would then continue a conversation about recent or future experiments, about new ideas and theories, and we never tired of enjoying the marvellous development of modern physics").[38] Here, the social was an extension of the scientific work. Marie stresses that Pierre's family was small and hers distant and that he was "very affectionate" ("très affectueux") to her family when they visited Paris or during the holidays.[39]

Marie elaborates still on the theme of family, a kind of typology or double image of private and public. In 1899, she says, Pierre travelled with her to Austrian Poland, in the Carpathians, where one of her sisters, a doctor, ran, with her husband, also a doctor, a sanatorium. Marie notes that Pierre wished to know all dear to her and wished to learn Polish and showed sympathy for Poland and its freedom. She came to know his thoughts and considered him much more than anything of which she dreamed at their union. His character impressed her:

> Constamment grandissait mon admiration pour ses qualités exceptionnelles, d'un-niveau si rare et si élevé, qu'il m'apparaissait parfois comme un être presque unique, par son détachement de toute vanité et de ces petitesses qu'on découvre chez soi-même et chez les autres, et que l'on juge avec indulgence, non sans aspirer à un idéal plus parfait. (My admiration for his exceptional qualities, of such a rare and high level, constantly grew that he sometimes appeared to me as an almost unique being, in his detachment from all vanity and from those pettinesses that we discover in ourselves and in others, and that we judge with indulgence, not without aspiring to a more perfect ideal).[40]

Marie looks back and sees what makes Pierre almost unique, detached from the vain and the petty. For her, Pierre's character is a focus. She sees this detachment as the secret of his "infinite charm" ("charme infini"), and she found "great attraction" in Pierre's "pensive face" ("figure pensive") and clear "gaze" ("regard"), and this impression was "increased by his kindness and the gentleness of his character" ("s'augmentait ensuite en raison de sa bienveillance et de la douceur de son caractère").[41] Marie says that Pierre did not feel combative and no one would not argue with him because he could not be angry – and here she quotes him on this point. He had few friends and no enemies because he was never even inadvertently hurtful. But, Marie observes, one could not convince him to deviate from an action, so that Pierre's father gave "him the name 'sweet stubborn' [gentle headstrong]" ("le nom de 'doux entêté'").[42]

Pierre's character continues to occupy Marie. She says that he expressed his opinions frankly and preferred that to diplomacy, that some thought him naïve, but she thinks he did this from will as opposed to instinct because his self-reflection allowed him to understand substance and the motives and thoughts of others. Marie adds that Pierre kept his judgement to himself until he made a decision. In science, he was not harsh or personal, and he took pleasure in any success. He said: "Il disait 'Qu'importe que je n'aie pas publié tel travail si un autre le publie', et pensait qu'en matière de science l'on doit s'intéresser aux choses et non aux personnes." ("What does it matter that I have not published a particular work if someone else publishes it?, and he believed that in matters of science, one should be interested in things and not in people.")[43] She stresses how he helped others and that his friendship, founded on a community of ideas, was rare, sure and faithful: "Plus rare encore a été le don de son affection, mais combien ce don a été complet envers son frère et envers moi-même !" ("Even rarer was the gift of his affection, but how complete was this gift to his brother and to myself!").[44] Marie emphasizes friendship and feeling, how he abandoned his reserve for trust and harmony. She reports Pierre's words when they were apart and then his view of work: "Quoi qu'il arrive, et dût-on être comme un corps sans âme, il faudrait travailler tout de même" ("Whatever happens, and even if we are like a body without a soul, we must work all the same").[45] Pierre is devoted to work even when it is soulless.

In Chapter 5, Marie discusses the discovery of radium as a dream come true. She describes again that in 1897, Pierre Curie was working on crystal growth and Marie had finished her study of the magnetization of hardened steels and notes the birth of their daughter, Irène, in September, after which Marie resumed work in the laboratory and intended to do a doctoral thesis. Marie recounts Roengen's discovery of X-rays and a phenomenon discovered in 1896 by Henri Becquerel, who studied uranium salts and came upon something different from what he was looking for – radioactivity. Marie decided to work in this area. Pierre found a place to set up experiments, and they used a quantitative method. They looked at ionization, which occurs with X-rays. As part of her method, Marie measured the weak currents passed through air that uranium rays ionize and drew on the methodology of Pierre and Jacques Curie. Pierre gave up his work on crystals and joined Marie in working on radioactivity. Marie goes into detail about the work and observes: "Nous avons pu reconnaître bientôt que la radioactivité se concentrait principalement dans deux fractions chimiques différentes, et nous avons été amenés à caractériser dans la pechblende la présence d'au moins deux radioéléments nouveaux : le *polonium* et le *radium*" ("We were soon able to recognize that the radioactivity was concentrated mainly in two different chemical fractions, and we were led to characterize the presence of at least two new radioelements in pitchblende: *polonium* and *radium*").[46] Marie reiterates the lack of resources for their work, and she notes the publications that Pierre and she produced in 1899 and 1900. She concludes the chapter with an enlightening description:

> En dehors des relations avec nos collaborateurs, nous ne voyions que peu de personnes au laboratoire parmi les physiciens et les chimistes, l'un ou l'autre venait de temps en temps, soit pour voir nos expériences, soit pour demander quelque conseil ou quelque renseignement à Pierre Curie, dont la compétence dans plusieurs branches de physique était bien connue. C'étaient alors des conversations devant le tableau noir, de celles dont on conserve un excellent souvenir parce qu'elles agissent comme un stimulant sur l'intérêt scientifique et sur l'ardeur au travail, sans interrompre le cours des réflexions et sans troubler cette atmosphère de paix et de recueillement qui est la véritable atmosphère d'un laboratoire.
>
> Aside from interacting with our colleagues, we saw only a few people in the laboratory among the physicists and chemists, one or the other came from time to time, either to see our experiments or to ask for advice or information from Pierre Curie, whose competence in several branches of physics was well known. These were conversations in front of the blackboard, the kind that we conserve as an excellent memory because they act as a stimulant to scientific interest and ardour for work, without interrupting the course of reflection and without disturbing the atmosphere of peace and contemplation which is the true atmosphere of a laboratory.[47]

Marie's description involves collaboration in the laboratory, in France and beyond. Cooperation and a community of interest – and not simply the work of Pierre and Marie Curie. Scientific discovery is individual but also collective.

In Chapter 6, Marie Curie speaks about resources at work, the burden of fame and the lateness of state support. She says that Pierre had no illusions about being appointed to a key professorship in Paris: "N'ayant passé ni par normale, ni par l'Ecole polytechnique, il manquait de l'appui souvent décisif que ces grandes Ecoles donnent à leurs élèves" ("Having attended neither the l'Ecole Normale nor the École Polytechnique, he lacked the often decisive support that these great schools give their students").[48] In 1900, Pierre Curie was offered a more junior six-month appointment at l'École Polytechnique as well as the Chair of Physics at the University of Geneva (including a laboratory for Marie), but despite the good conditions in Geneva, they decided to remain in Paris. A Chair in Physics at the Sorbonne was vacant, and thanks to Henri Poincaré, he was appointed a lecturer, and Marie was in charge of lectures in Physics at the École Normale Supérieure for Young Women in Sèvres. Among the many challenges owing to lack of space, time and resources, Marie notes one outside the university: "D'ailleurs notre travail ne pouvait plus progresser qu'avec l'aide de moyens industriels pour le traitement de la matière première" ("Moreover, our work could no longer progress without the help of industrial means for processing the raw material").[49] Profit was not the motive of Pierre and Marie:

> D'accord avec moi, il renonça à tirer un profit matériel de notre découverte : en conséquence, nous n'avons pris aucun brevet et nous avons publié, sans aucune réserve, tous les résultats de nos recherches, ainsi que les procédés de préparation du radium (In agreement with me, he renounced to take a material profit from our discovery: consequently, we did not take out any patents and we published, without reservation, all the results of our research, as well as the processes for the preparation of radium).[50]

Marie, who says she presented her doctoral thesis in 1903, gives a detailed and lucid account of the scientific research and work conditions as well as results: "L'activité du radium pur a été reconnue supérieure à toutes nos prévisions. À poids égal, cette substance émet un rayonnement plus d'un million de fois plus intense que l'uranium" ("The activity of pure radium was recognized as superior to all our predictions. Weight for weight, this substance emits radiation more than a million times more intense than uranium").[51] Part of Marie's narrative is an account of Pierre's overwork: "La fatigue physique due aux nombreuses courses auxquelles il était obligé lui était d'autant plus pénible qu'il souffrait de crises de douleurs aigues, rendues de plus en plus fréquentes par le surmenage" ("The physical fatigue owing to the numerous errands he was required to undertake was all the more painful as he suffered from attacks of acute pain, made increasingly frequent by overwork").[52]

Among other subjects, she discusses publications, which are part of Pierre's work:

> Toutes ces recherches sur la radioactivité sont fondamentaleset s'adressent à des sujets très variés. Plusieurs ont pour but l'étude de l'émanation, ce corps gazeux étrange que produit le radium, et qui est responsable, pour une forte part, du rayonnement intense, communément attribué à ce

> dernier. Pierre Curie mit en évidence, dans une étude approfondie, la loi rigoureuse et invariable, suivant laquelle l'émanation se détruit quelles que soient les conditions dans lesquelles elle se trouve.
>
> Many of these studies focus on the study of emanation, the strange gaseous substance produced by radium, which is largely responsible for the intense radiation commonly attributed to it. Pierre Curie demonstrated, in an in-depth study, the rigorous and invariable law according to which the emanation is destroyed regardless of the conditions in which it is found.[53]

Marie analyses science in and of itself and in relation to its environment. She is frank about the challenges and does not separate science from work, describing less than ideal circumstances. Moreover, Marie signals what she considers to be important: "Plus frappante encore a été la découverte du dégagement de chaleur du radium" ("Even more striking was the discovery of the release of heat from radium").[54] She discusses the related research in France and in other countries, most notably in Britain, and she sums up:

> Cette succession rapide de découvertes bouleversant les conceptions scientifiques acquises par la physique et la chimie, n'a pas été sans rencontrer tout d'abord des doutes et de l'incrédulité, mais une grande partie du monde scientifique l'accueillit avec enthousisame [**enthousiasme**]. En même temps, la notoriété de Pierre Curie grandit en France et à l'étranger (This rapid succession of discoveries, overturning scientific conceptions acquired through physics and chemistry, was not without initial doubt and disbelief, but a large part of the scientific world welcomed it with enthusiasm. At the same time, the reputation of Pierre Curie grew in France and abroad).[55]

In 1903, the work of Marie and Pierre came to be recognized in London, where they saw Lord Kelvin: ("A few months later, the Davy Medal was awarded to him (jointly with me) by the Royal Society of London, and almost at the same time, we were jointly awarded the Nobel Prize in Physics with Henri Becquerel") ("la médaille Davy lui était décernée (en commun avec moi) par la Société Royale de Londres, et presque en même temps nous obtenions, en commun avec Henri Becquerel, le prix Nobel de physique").[56] Marie says that their health prevented Pierre and her from attending the award ceremony in December 1903 and that they went to Stockholm in June 1905 when Pierre Curie delivered his Nobel Lecture. Although Marie outlines the benefits of the Nobel, the recognition and the funds that came with it, she also described the burden of fame for Pierre: "Toutefois la publicité déterminée par cet heureux événement pesa aussitôt très lourdement sur un homme qui n'y était ni préparé, ni habitué" ("However, the publicity generated by this happy event immediately weighed heavily on a man who was neither prepared nor habituated to it").[57] Despite the benefits, Pierre still had some concerns beyond the fame, which he expresses in a letter to E. Gouy on 22 January 1904: "Ils créent une chaire sans programme, ce sera quelque chose comme un cours au Collège de France, et je crois que je serai obligé de changer de sujet chaque année ce qui me donnera beaucoup de mal" ("They are creating a chair without a programme; it will be something like a course at the Collège de France, and I believe I will be forced to change subjects every year, which will give me a lot of trouble").[58] The workload continues to worry Pierre. In another letter to Gouy on 31 January 1905, as Marie quotes it, Pierre says that Marie and he had to give up going to the Swedish Academy and admits that they are worn out: "A la vërité, je n'arrive à maintenir me en état qu'en évitant toute fatigue physique. Ma femme est dans le même cas que moi, et il ne faut plus songer aux grandes journées de travail d'autrefois" ("In truth, I can only maintain my fitness by avoiding all physical fatigue. My wife is in the same situation as me, and we must no longer think about the long workdays of the past.").[59] Years of hard work have led to exhaustion. Marie also quotes from a letter from Pierre to Gouy on 24 July 1905 about a similar theme: "Voilà plus d'un an que je n'ai fait aucun travail, et je n'ai pas un moment à moi" ("It has been over a year since I have done any work, and I do not have a moment to myself") and he then says: "Evidemment, je n'ai pas encore trouvé le moyen de nous défendre contre l'émiettement de notre temps, et c'est cependant bien nécessaire. C'est une question de vie ou de mort au point de vue intellectuel" ("Evidently, I have not yet found a way to defend ourselves against the crumbling of our time, and yet it is very necessary. It is a matter of life or death from an intellectuel point of view"). Pierre is frustrated over not being able to work, it seems, on his scientific research. Fame had crumbled the time of Pierre and Marie Curie. She frames and interprets what he is saying in the letters and how their family and professional lives were, how Pierre was with their daughters, how a professorship was created for him in Paris, but it did not have a laboratory attached to it. Marie notes that Pierre stood firm:

> Sa fermeté obtint gain de cause; la création de la chaire nouvelle fut complétée par l'attribution d'un crédit de laboratoire et d'un personnel pour le nouveau service (un chef de travaux, un préparateur, un garçon de laboratoire). La situation de chef de travaux m'était offerte, ce qui était aussi une grande satisfaction pour Pierre Curie (His firmness was successful; the creation of the new chair was completed by the allocation of a laboratory credit and staff for the new department (a laboratory chief, a preparator, and a laboratory boy). I was offered

the position of laboratory chief [chief of works], which was also a great satisfaction for Pierre Curie).[60]

Pierre seems to have to had to struggle for good conditions for Marie and him even after their Nobel Prize. Marie represents the situation of her husband and their young daughters. Marie humanizes science. The struggles and life of her family are an important part of her account. Marie has a talent for expressing feelings and atmosphere, mixing memories with analysis, sometimes moving back and forth in time. Marie shifts ahead to the time, after the death of Pierre, when the hangar where they did their work was demolished:

> Plus tard, il a fallu le détruire pour faire place au nouveau bâtiment de l'Ecole de physique, mais des photographies en ont été conservées. Prévenue par le fidèle Petit, j'y fis, seule, hélas! un dernier pélerinage: au tableau noir était encore restéel écriture de celui qui avait été l'âme de ce lieu; l'humble asile de son travail était tout imprégné de son souvenir. La cruelle réalité semblait un mauvais rêve; on espérait presque voir paraître la haute silhouette et entendre résonner la voix familière[.]
>
> Later, it had to be demolished to make way for the new Physics School building, but photographs of it have been preserved. Warned by the faithful Petit, I made, alone, alas! a final pilgrimage: on the blackboard there still remained the writing of the one who had been the soul of this place; the humble refuge of his work was imbued with his memory. The cruel reality seemed like a bad dream; one almost hoped to see the tall silhouette appear and hear the familiar voice resonate.[61]

The ghost of Pierre is in Marie's bones, in her words. Her love for him she expresses here and in moments throughout the biography. Here, she gets to his death before his death in the larger narrative. This inset or almost ekphrasis, this narrative within a narrative, plays with the time, with a trace before she comes to the death of Pierre in her chronology of his life.

Another passage involves her reflections on the example of Pierre Curie after his death:

> On ne peut s'empêcher d'éprouvé quelque amertume à la pensée que cette concession a été la dernière, et qu'en définitive un des premiers savants français n'eut jamais à sa disposition un laboratoire convenable, alors que cependant son génie s'était révélé dés l'âge de vingt ans. Sans doute, s'il eût vécu plus longtemps, il eût bénéficié tôt ou tard de conditions de travail satisfaisantes, mais lors de son décès prématuré, à quarante-huit ans. Il en était encore dépourvu. Imagine-t-on le regret de l'ouvrier enthousiaste et désintéressé d'une grande oeuvre, retardé

dans la réalisation de son rêve parle manque constant de moyens ? Et pouvons-nous songer sans un sentiment de peine profonde au gaspillage, irreparable entre tous, du plus grand bien de la nation le génie, les forces et le courage de ses meilleurs enfants.

One cannot help feeling some bitterness at the thought that this concession was the last, and that ultimately one of the first [foremost] French scientists never had a suitable laboratory at his disposal, even though his genius had been revealed at the age of twenty. No doubt, if he had lived longer, he would have benefited sooner or later from satisfactory working conditions, but at the time of his premature death, at forty-eight, he was still deprived of them. Can one imagine the regret of the enthusiastic and disinterested worker of a great work, delayed in the realization of his dream by a constant lack of means? And can we think without a feeling of deep sorrow of the waste, irreparable above all, of the greatest good of the nation, the genius, the strength and the courage of its best children.[62]

The lack of resources is sorrowful. Marie speaks of Pierre's "genius" ("génie") – a word she repeats twice in this passage. She appeals to imagination to have someone – generally but the reader who reads this – imagine what regret Pierre felt owing to the delay that the inadequate means caused. For Marie, the private and the public meet in this what if because how Pierre was treated affected him but also the nation. This is also an implied lesson about the necessity of funding science well. Pierre and the nation lost because of the delay. This is a remembrance of time lost.

Returning to the chronology, Marie discusses Pierre's professorship and teaching on symmetry, the physics of crystals and related topics as well as lectures on radioactivity, whose discovery revolutionized science. She relates this experience to his research: "Très préoccupé de la préparation de son cours et souvent souffrant, Pierre Curie continuait cependant à travailler au laboratoire, dont l'organisation se faisait progressivement" ("Very preoccupied with the preparation of his courses and often suffering [illness], Pierre Curie nevertheless continued to work in the laboratory, whose organization was been made progressively").[63] At length, Marie assesses Pierre's gifts and contributions, a selection of which I quote:

Ses facultés intellectuelles étaient alors en plein développement ; on pouvait admirer la sûreté et la rigueur de ses raisonnements sur les théories de la physique, sa compréhension claire des principes fondamentaux et un certain sens profond des phénomènes qu'il avait d'instinct, mais qu'il avait formé et perfectionné au cours de toute une vie consacrée à la recherche et à la réflexion. Son habileté expérimentale, remarquable des ses débuts, s'était augmentée par la pratique. Il éprouvait un plaisir d'artiste à réaliser un montage délicat. Il se plaisait aussi à imaginer et à construire des appareils nouveaux, et il m'arrivait de lui dire en plaisantant qu'il n'aurait pu être heureux sans faire au moins un essai de ce genre tous les six mois. Sa curiosité naturelle et la vivacité de son imagination le pousaient à s'engager dans des directions très variées ; il pouvait changer d'objet de recherche avec une souplesse surprenante.

Il était extrêmement soucieux de probité scientifique et d'une entière correction dans ses publications. Celles-ci, très parfaites dans leur forme, ne l'étaient pas moins en ce'qui concerne l'esprit critique appliqué à soi-même, et la volonté de ne rien affirmer qui ne parût entièrement clair.

His intellectual faculties were then in full development; one could admire the sureness and rigour of his reasoning on the theories of physics, his clear understanding of fundamental principles, and a certain profound sense of phenomena that he possessed instinctively, but which he had formed and perfected over the course of a lifetime devoted to research and reflection. His experimental skill, remarkable from the beginning, had increased through practice. He took an artistic pleasure in creating a delicate setup. He also enjoyed imagining and constructing new apparatus, and I sometimes jokingly told him that he could not have been happy without conducting at least one such experiment every six months. His natural curiosity and the liveliness of his imagination pushed him to pursue a wide variety of directions; he could change the subject of his research with surprising flexibility.

He was extremely concerned with scientific integrity and complete correctness in his publications. These, very perfect in their form, were no less so in what concerned the critical spirit applied to oneself, and the will to affirm nothing which did not appear entirely clear.[64]

Marie stresses that Pierre was at the height in theory and practice, praising his reason, clarity, understanding, skill in experiments, artistry, imagination, curiosity, flexibility in research, integrity and critical spirit. In this portrait, Pierre is accomplished and ethical in theory and experimentation. She then quotes Pierre on hypotheses to amplify this verbal portrait and comments herself after the quotation. Her interpretation dovetails her words and his words. Marie also speaks about the improvements Pierre made to his teaching. She quotes him about his view that science be the dominant teaching in high schools for boys and girls, but he thinks that a motion to that effect will have little chance of success. She also comes to her reflection on his career before his death:

Cette dernière période si féconde de sa vie devait, hélas, avoir une fin prochaine. Son admirable carrière scientifique a été brusquement brisée au moment même où il pouvait espérer que les années de travail à venir seraient moins dures que celles qui les avaient précédées (This last, very fruitful period of his life was, alas, to come to an imminent end. His admirable scientific career was abruptly broken at the very moment when he could have hoped that the years of work to come would be less hard than those that had preceded them).[65]

Here, is the lead-up to a description of a life cut short, to the gruesome death itself. And to Pierre's shortened life and sudden death we turn in the next chapter.

Notes

1 My translation; Marie Curie, *Pierre Curie* (Les grands hommes de France, collection dirigée par Paul Gautier) (Paris: Payot, 1924), 26; see Marie Curie, *Pierre Curie*, trans. Charlotte Kellogg and Vernon L. Kellogg, With an Introduction by Mrs. William Brown Meloney and Autobiographical Notes by Marie Curie (New York: The Macmillan Company, 1923), 51.
2 My paraphrase; Curie, *Pierre Curie* (Les grands ...), 28; see Curie, *Pierre Curie*, 51.
3 *Pierre Curie* (Les grands ...), 29; my translation; see Curie, *Pierre Curie*, 52.
4 *Pierre Curie* (Les grands ...), 29; my translation; see Curie, *Pierre Curie*, 53.
5 My translation; *Pierre Curie* (Les grands ...), 30; see Curie, *Pierre Curie*, 54.
6 My translation; *Pierre Curie* (Les grands ...), 30; see Curie, *Pierre Curie*, 54.
7 My translation; *Pierre Curie* (Les grands ...), 30–31; see Curie, *Pierre Curie*, 54–55.
8 My translation; *Pierre Curie* (Les grands ...), 31; see Curie, *Pierre Curie*, 55–56.
9 *Pierre Curie* (Les grands ...), 36; my translation; see Curie, *Pierre Curie*, 62.
10 My translation; *Pierre Curie* (Les grands ...), 36–37; see Curie, *Pierre Curie*, 54–55.
11 My translation; *Pierre Curie* (Les grands ...), 37; see Curie, *Pierre Curie*, 63.
12 My translation; *Pierre Curie* (Les grands ...), 38; see Curie, *Pierre Curie*, 65.
13 My translation; *Pierre Curie* (Les grands ...), 39; see Curie, *Pierre Curie*, 65.
14 My translation; *Pierre Curie* (Les grands ...), 39–40; see Curie, *Pierre Curie*, 65.
15 My translation; *Pierre Curie* (Les grands ...), 40; see Curie, *Pierre Curie*, 66.
16 My translation; *Pierre Curie* (Les grands ...), 40; see Curie, *Pierre Curie*, 66.
17 *Pierre Curie* (Les grands ...), 41; my translation; see Curie, *Pierre Curie*, 67.
18 *Pierre Curie* (Les grands ...), 41; my translation; see Curie, *Pierre Curie*, 67.
19 My translation; *Pierre Curie* (Les grands ...), 43; see Curie, *Pierre Curie*, 66.
20 My translation; *Pierre Curie* (Les grands ...), 47; see 46; see Curie, *Pierre Curie*, 74; see 73.
21 *Pierre Curie* (Les grands ...), 47; my translation; see Curie, *Pierre Curie*, 74.
22 My translation; *Pierre Curie* (Les grands ...), 47; see Curie, *Pierre Curie*, 74.
23 *Pierre Curie* (Les grands ...), 47; my translation; see Curie, *Pierre Curie*, 75.
24 *Pierre Curie* (Les grands ...), 48, see 47; my translation; see Curie, *Pierre Curie*, 75.
25 My translation: *Pierre Curie* (Les grands ...), 48; see Curie, *Pierre Curie*, 75.
26 *Pierre Curie* (Les grands ...), 48; my translation; see Curie, *Pierre Curie*, 75.
27 *Pierre Curie* (Les grands ...), 48; my translation; see Curie, *Pierre Curie*, 76.
28 *Pierre Curie* (Les grands ...), 48–49; my translation; see Curie, *Pierre Curie*, 76.

29 *Pierre Curie* (Les grands ...), 49; my translation; see Curie, *Pierre Curie*, 76.
30 *Pierre Curie* (Les grands ...), 50; my translation; see Curie, *Pierre Curie*, 78, see 77.
31 *Pierre Curie* (Les grands ...), 50; my translation; see Curie, *Pierre Curie*, 78.
32 *Pierre Curie* (Les grands ...), 51; my translation; see Curie, *Pierre Curie*, 78–79.
33 *Pierre Curie* (Les grands ...), 51; my translation; see Curie, *Pierre Curie*, 78–79.
34 My translation; *Pierre Curie* (Les grands ...), 51; see *Pierre Curie*, 79.
35 *Pierre Curie* (Les grands ...), 54; my translation; see Curie, *Pierre Curie*, 82.
36 *Pierre Curie* (Les grands ...), 55; my translation; see Curie, *Pierre Curie*, 83.
37 *Pierre Curie* (Les grands ...), 55; my translation; see Curie, *Pierre Curie*, 83.
38 *Pierre Curie* (Les grands ...), 58; my translation; see Curie, *Pierre Curie*, 87.
39 My translation; *Pierre Curie* (Les grands ...), 58–59; see Curie, *Pierre Curie*, 88.
40 *Pierre Curie* (Les grands ...), 59; my translation; see Curie, *Pierre Curie*, 88–89.
41 *Pierre Curie* (Les grands ...), 59; my translation; see Curie, *Pierre Curie*, 89.
42 *Pierre Curie* (Les grands ...), 60; my translation; see Curie, *Pierre Curie*, 89.
43 *Pierre Curie* (Les grands ...), 60–61; my translation; see Curie, *Pierre Curie*, 90.
44 *Pierre Curie* (Les grands ...), 61; my translation; see Curie, *Pierre Curie*, 90.
45 *Pierre Curie* (Les grands ...), 62; my translation; see Curie, *Pierre Curie*, 92.
46 *Pierre Curie* (Les grands ...), 67; my translation; see Curie, *Pierre Curie*, 98.
47 *Pierre Curie* (Les grands ...), 74; my translation; see Curie, *Pierre Curie*, 106.
48 *Pierre Curie* (Les grands ...), 75; my translation; see Curie, *Pierre Curie*, 107. There is no accent on the E in l'Ecole in the original French text.
49 *Pierre Curie* (Les grands ...), 77; my translation; see Curie, *Pierre Curie*, 110.
50 *Pierre Curie* (Les grands ...), 78; my translation; see Curie, *Pierre Curie*, 110.
51 *Pierre Curie* (Les grands ...), 80; my translation; see Curie, *Pierre Curie*, 114.
52 *Pierre Curie* (Les grands ...), 81; my translation; see Curie, *Pierre Curie*, 114–115.
53 *Pierre Curie* (Les grands ...), 82; my translation; see Curie, *Pierre Curie*, 116.
54 *Pierre Curie* (Les grands ...), 83; my translation; see Curie, *Pierre Curie*, 117.
55 *Pierre Curie* (Les grands ...), 88; my translation; see Curie, *Pierre Curie*, 123. I have corrected the misprint –it should be *enthousime* – in the original book.
56 *Pierre Curie* (Les grands ...), 89; my translation; see Curie, *Pierre Curie*, 124–125.
57 *Pierre Curie* (Les grands ...), 90; my translation; see Curie, *Pierre Curie*, 125.
58 *Pierre Curie* (Les grands ...), 91; my translation; see Curie, *Pierre Curie*, 127.
59 *Pierre Curie* (Les grands ...), 92; my translation; see Curie, *Pierre Curie*, 127.
60 *Pierre Curie* (Les grands ...), 94; my translation; see Curie, *Pierre Curie*, 130–131.
61 *Pierre Curie* (Les grands ...), 94; my translation; see Curie, *Pierre Curie*, 131.
62 *Pierre Curie* (Les grands ...), 95; my translation; see Curie, *Pierre Curie*, 132.
63 *Pierre Curie* (Les grands ...), 96; my translation; see Curie, *Pierre Curie*, 134.
64 *Pierre Curie* (Les grands ...), 96–97; my translation; see Curie, *Pierre Curie*, 134–135.
65 *Pierre Curie* (Les grands ...), 98–99; my translation; see Curie, *Pierre Curie*, 136.

8 Marie Curie

The Death of Pierre Curie and the Autobiography

The death of Pierre, the tributes to him and Marie's autobiography occupy this chapter. Here, Marie mourns and she takes up the weight of the reputation of her husband and herself as well as furthering science. The emphasis shifts from representing Pierre and the Curies to her self-representation.

Marie turns to the events of 1906, when Pierre, unwell and tired, spent Easter in the Chevreuse Valley with Marie and their two daughters with whom he played in the meadow and talked with Marie about them. Pierre returned to Paris, where, at the Physics Society, he had a long conversation with Henri Poincaré, about teaching. As Marie and he walked home, he talked about the culture he imagined and "was happy to feel that his feelings were shared by me" ("heureux de sentir que son sentiment était partagé par moi").[1] Marie shifts to the next day, 19 April 1906, when Pierre attended a meeting, after which, when crossing Rue Dauphine, he was hit by a truck [a horse and wagon] coming from the Pont Neuf and fell under its wheels.[2] Marie's retrospective reflection on the death is touching:

> La contusion à la tête fut instantanément mortelle, et ainsi fut détruite l'espérance que l'on pouvait fonder sur l'être merveilleux qui venait de disparaître. Dans le cabinet de travail où il ne devait plus revenir, les renoncules d'eau, qu'il avait rapportées de la campagne, étaient toutes fraîches encore. (The contusion to his head was instantly fatal, and thus was destroyed any hope that could have been placed in the marvellous being who had just disappeared. In the work room [study], to which he was not again to return, the water buttercups he had brought back from the countryside were fresh still).[3]

This juxtaposition of the sudden and violent accident and the plant he picked (as he used to pick with his brother Jacques in their youth) in the country is wistful because, as Marie says, he did not return to them. Pierre left behind his wife and two young daughters. One of the daughters, Eve,

DOI: 10.4324/9781003651871-9

wrote a biography of her mother, with a whole chapter on the fatal day of her father's death. I quote the English edition published in 1938 as it is more accessible for the reader when she quotes her grandfather, Pierre's father, when Paul Appell was there to tell him of his son's death: "'What was he dreaming of this time?'"[4] When Marie returned home, she was told and her daughter reports her words as "'Pierre is dead? Dead? Absolutely dead?'"[5] This report is years later by someone who was a young child, so it is difficult, years after her grandfather had died and a few years or so after her mother had died to verify, but her own interpretation of her mother remains and is hers: "Mme Curie, on that day in April, became not only a widow, but at the same time a pitiful and incurably lonely woman."[6] But I get ahead of the story as Marie does. The Curies are a highly literate and educated family, and Marie interprets Pierre and herself, and Eve interprets her. There is a mix of autobiography and biographies over time, and readers interpreting the interpretations.

In Chapter 7, Marie discusses the mourning of France, the laboratories and related matters. She notes that her story has shown what Pierre was for his father, brother and wife, but that the daughters were too young to understand the misfortune of losing a devoted and tender father, something Marie and Pierre's father tried to mitigate so the disaster would not overshadow their childhood. She tells of the burial at Sceaux in the family plot, and indeed Pierre and Marie and Pierre's parents, Sophie and Eugène, were buried in a simple burial there until 1995 when Pierre and Marie were reinterred at the Pantheon, where we can see this resting place to this day. Jacques told Marie of the great qualities and uniqueness of Pierre. Marie says that to ensure that Pierre's work continued, the Faculty of Sciences of Paris offered her Pierre's chair. She said she hoped to build the laboratory Pierre never had. And she noted that this has been partially realized. The street by which the institute was reached was renamed after Pierre Curie. Marie does not miss an opportunity to advocate for new resources:

> C'est maintenant une nécessité reconnue par les personnes les plus autorisées que la France doit posséder un Institut du Radium comparable à ceux qui ont été fondés en Angleterre et en Amérique pour appliquer la curiethérapie, qui est devenue un moyen efficace de lutte contre le cancer. (It is now a necessity recognized by the most authoritative people that France must possess a Radium Institute comparable to those founded in England and America to apply curietherapy [brachytherapy], which has become an effective means of fighting against cancer)[7]

She also records that to honour Pierre, the French Physical Society decided to publish his complete work in a volume edited by Paul Langevin and published in 1908 with a preface Marie wrote. She praises Pierre's classical form while she describes the contents of the volume while expressing regret:

> On peut regretter que Pierre Curie n'ait pas utilisé ses qualités de savant et d'écrivain pour la rédaction de mémoires étendus ou de livres. Ce n'est pas le désir de le faire qui lui manquait ; il avait plusieurs projets de ce genre qui lui étaient chers. (We can regret that Pierre Curie did not use his talents as a scholar and writer to write extensive memoirs or books. It was not the desire to do so that he lacked; he had several projects of this kind that were dear to him).[8]

She adds that Pierre was not able to carry out these projects owing to the longstanding hardships against which he struggled at work. She tries to fill the gap.

Coming to the end of the biography of Pierre, Marie provides the lesson:

> Et maintenant, jetons un regard sur l'ensemble de ce récit, on j'ai tenté d'évoquer l'image d'un homme qui, attaché inflexiblement au service de son idéal, a honoré l'humanité par une existence de travail vécue, dans le silence, dans la simple grandeur de son génie et de son caractère. (And now, let us take a look at the whole of this story, in which I have tried to evoke the image of a man who, inflexibly attached to the service of his ideal, honoured humanity with a life of work lived in silence, in the simple grandeur of his genius and character).[9]

This sentence is a summa of the biography, a portrait in miniature.

Marie Curie is a skilful writer, and the passages here towards the end of the biography are exemplary, one of which I quote here at some length:

> Il avait la foi de ceux qui ouvrent des voies nouvelles il savait qu'il avait une haute mission à remplir, et le rêve mystique de sa jeunesse le poussait invinciblement, en dehors du chemin usuel de la vie, dans une voie qu'il nommait antinaturelle, car elle signifait le renoncement à la douceur de l'existence. Pourtant, résolument, il subordonna à ce rêve ses pensées et ses désirs il s'y adapta et s'y identifia de manière de plus en plus complète. Ne croyant qu'à la puissance pacifique de la science et de la raison, il vécut pour la recherche de la vérité. Sans préjugé et sans parti pris, a apporta la même loyauté dans l'étude des choses que dans la compréhension des autres hommes et de lui-même. Détaché de toute passion commune, ne cherchant ni la suprématie, ni les honneurs, il n'avait point d'ennemis, bien que l'effort accompli sur lui-même en ait fait un de ces êtres d'élite, que l'on trouve à toutes les 'époques de

la civilisation en avance sur leur temps. Et ainsi que ceux-là, il pouvait exercer une influence profonde, par le seul rayonnement de sa puissance intérieure.

Il est utile de comprendre combien une pareille existence représente de sacrifice. La vie du grand savant dans son laboratoire n'est pas comme beaucoup peuvent le croire une idylle paisible ; elle est plus souvent une lutte opiniâtre livrée aux choses, à l'entourage et surtout à soi-même. Une grande découverte ne jaillit pas du cerveau du savant tout achevée, comme Minerve surgît tout équipée de la tête de Jupiter ; elle est le fruit d'un labeur préliminaire accumulé. Entre des journées de production féconde viennent s'intercaler des journées d'incertitude où rien resemble réussir, ou la matièreelle-même semble hostile, et c'est alors qu'il faut résister au découragement. Et sans jamais se départir de sa patience inlassable, Pierre Curie me disait parfois : "Elle est pourtant dure, la vie que nous avons choisie."

He had the faith of those who open new paths; he knew he had a high mission to fulfil, and the mystical dream of his youth pushed him irresistibly away from the usual path of life, toward a path he called unnatural, for it meant renouncing the sweetness of existence. Yet, resolutely, he subordinated his thoughts and desires to this dream; he adapted and identified with it more and more completely. Believing only in the peaceful power of science and reason, he lived for the search for truth. Without prejudice or bias, he brought the same loyalty to the study of things as he did to understanding other men and himself. Detached from all common passions, seeking neither supremacy nor honours, he had no enemies, although the effort he made over himself made him one of those elite beings found in all eras of civilization ahead of their time. And like them, he could exert a profound influence, simply by radiating his inner power.

It is useful to understand how much sacrifice such an existence represents. The life of a great scientist in his laboratory is not, as many might believe, a peaceful dream; it is more often a stubborn struggle waged against things, against those around him, and above all against oneself. A great discovery does not spring from the scientist's brain fully completed, like Minerva emerging fully equipped from the head of Jupiter; it is the fruit of accumulated preliminary labour. Between days of fruitful production come days of uncertainty when nothing seems to succeed, or the material itself seems hostile, and it is then that one must resist discouragement. And without ever losing his tireless patience, Pierre Curie sometimes said to me, "It is hard, however, the life we have chosen."[10]

This passage amplifies many of the themes Marie has discussed in the biography of Pierre. She reiterates his mystical dream of science from his youth, which was against nature because it renounced the sweetness of life and involved a discipline that subordinated his thoughts and feelings to this dream. Pierre chose the peace of science and reason. In Marie's portrait, he focused on understanding things, people and himself. This focus meant that, with an inner strength, he strove for an understanding ahead of his time and not for honours. Marie continues to stress that a great scientist faces struggles in the laboratory, and a great discovery means great labour, is not some kind of birth as in Greek mythology of Minerva from the head of Jupiter. Marie's style has some literary flourishes. The scientist sometimes faces unsureness and hostile materials but must not be discouraged. Marie shows the bond Pierre shared with her by quoting his words to her about the difficult life they had chosen. All this encapsulates her earlier portrait and performs a repetition. Marie hammers away at the image of Pierre. She is a guardian of his spirit, of his reputation. He is her mentor, husband, co-investigator and an exemplary scientist. For Marie, not only does she grieve for Pierre, but she also sees him as a model scientist and believes that his life is an argument for science and devotion to it.

Marie augments this lesson by generalizing from Pierre Curie to scientists, who give themselves and render service, but, like Pierre, they do not have what they need and are not secure, exhausting, through worry, youth and strength in society given to luxury and wealth that does not value science, part of its moral heritage, improving life and reducing suffering. She adds that neither public nor private sources currently support science and scientists effectively. Once more, Marie invokes Pasteur on scientific discoveries and people wanting their country to invest in laboratories, where nature is read and progress is made as opposed to the barbaric, fanatical and destructive forces. Marie begins a formal wish in a subjunctive "Puisse cette vérité" ("May this truth") to express a wish, to exhort, to hope, to pray, that this truth be disseminated in public opinion to create a good future.[11]

Marie adds a section of opinions on Pierre Curie. She chooses extracts from different publications of eminent scientists witnessing and remembering Pierre and begins by quoting Henri Poincaré from the Proceedings of the Academy of Sciences in December 1906. Poincaré represents Pierre Curie as someone on whom science and France could rely, of accomplishment and promise. He tells about the evening before Pierre's death when they talked about Pierre's projects and ideas, how Poincaré admired his thought, his original and lucid mind and then how the next day showed the blind forces that crush us. Poincaré speaks of the loss among friends, colleagues, scholars in France and beyond. For instance, Poincaré says:

> Curie apportait, dans l'étude des phénomènes physiques, je ne sais quel sens très fin qui, lui faisant deviner les analogies insoupçonnées, lui permettait de s'orienterà travers un dédale de complexes apparences

> où d'autres se seraient égares.... Les vrais physiciens, comme Curie, ne regardent ni en dedans d'eux-mêmes, ni à la surface des choses, ils savent voir sous les choses. (Curie brought to the study of physical phenomena an indescribable sense which, allowing him to divine unsuspected analogies, allowed him to orient himself through a maze of complex appearances where others would have lost their way.... True physicists, like Curie, look neither inward nor at the surface of things; they know how to see beneath things).[12]

Here is the image of a brilliant physicist, someone who sees what most cannot. Poincaré says that Pierre Curie was pleasant, charming, modest, gentle, uncompromising and, above all, sincere: he rendered sincerity to science, an example of duty while showing a love of truth and faith in science.

Marie also includes a notice by N.-D. Gernez, of the Institut de France. Gernez says of the life of Pierre Curie that it is "a life so rich in brilliant disappointments and brilliant insights that it soon earned universal admiration" ("vie si riche en découvertes brillantes et en vues géniales qu'elle lui a valu bientôt une admiration universelle") while observing that

> he was, and will remain, a remarkable figure among all in the scientific history of our time; his contemporaries find in him a precious example of a dedication to science that was both stubborn and selfless. There are few lives purer and more justly celebrated. (il était et il restera une figure remarquable entre toutes dans l'histoire scientifique de notre époque; ses contemporains trouvent en lui le précieux exemple d'un dévouement à la science à la fois opiniâtre et désintéressé. Il est peu de vies plus pures et plus justement célèbres).[13]

Gernez stresses Pierre's brilliance, purity and exemplary dedication to science. This testimony and remembrance amplifies, as Poincaré's comments do, the verbal portrait Marie Curie made in the body of her biography of Pierre Curie. In May 1906, Jean Perrin says that Pierre, a master and friend to all, died in his prime. Perrin stresses the example and genius of Pierre Curie, the sincere, free, audacious thought, his great soul and his character, intelligence, selflessness and kindness among other things. Perrin speaks on behalf of others as they honour the memory of Pierre, his face and the secret of radiation that improved others. The next testimony from which Marie Curie quotes is that of C. Chéneveau in April 1906. Chéneveau speaks of Pierre's immense kindness (bounty, generosity) ("immense bonté") and concludes: "I have never seen tears more sincere, nor more heart-rending, than those shed by his laboratory boys [workers] at the news of his sudden disappearance [death]" ("je n'ai jamais vu de pleurs plus sincères, ni plus déchirants, que ceux qui furent versés par ses garçons de laboratoire à l'annonce de sa brusque disparition").[14] Pierre Curie was kind to all in his laboratory.

The last testimony is that of Paul Langevin in July 1906. Among other things, this account represents memories of Pierre's love of science, his kind and thoughtful face, his work and simplicity, the apparatus he shaped, being his student, his blackboard, his curiosity, memories of him as a bouquet on his grave as a man of great character and thought. Langevin ends the tribute thus:

> Entièrement affranchi d'antiques servitudes, amoureux passionné de raison et de clarté, il a donné l'exemple, en prophète inspiré des vérités futures, de ce que peut réaliser en beauté morale et en bonté, dans une esprit libre et droit, le courage constant, la propriété mentale, de toujours repousser ce qu'il ne comprehends pas et de mettre sa vie d'accord avec ce rêve. (Entirely freed from ancient servitudes, a passionate lover of reason and clarity, he set an example, as an inspired prophet of future truths, of what can be achieved in moral beauty and goodness, in a free and upright mind, through constant courage, mental propriety, always pushing back against what he does not understand and putting his life in accord with this dream).[15]

The use of "repousser" is a little jarring; it could be translated as "rejected," but this usage makes it sound as though he rejected what he could not understand instead of another sense of the word, which means opposing successfully or not accepting that something is unknowable rather than pushing it away or pushing it back in time or delaying or being hostile to it, as other uses of "repousser" can mean.[16] I take Langevin to mean that Pierre Curie was curious and would try to understand the unknown.

At the end of the English version of this life of Pierre Curie, and not in the French version, Marie includes "Autobiographical Notes," and so I will examine this part of the publication briefly. Here are the opening and source of the idea for these notes: "I HAVE been asked by my American friends to write the story of my life. At first, the idea seemed alien to me, but I yielded to persuasion."[17] The idea came from her friends in the United States. Unlike the French version, the English version supplements Marie's biography of Pierre with her own autobiography, and so the story – the book – changes, the French version appearing with the copyright date of 1924 and the English version with the date of 1923. The English version also announces on its title page an Introduction by Mrs. William Brown Meloney as well as the autobiographical notes of Marie Curie, so that the life of Pierre is literally framed, the one coming after Marie's Preface and the other after the life of Pierre. Mrs. Meloney begins "THE LIFE STORY OF PIERRE CURIE" – her introduction – with Marie Curie:

> Every little while a man or a woman is born to serve in some big way. Such a one surely is Marie Curie. Her discovery of radium has advanced science, relieved human suffering and enriched the world. The spirit in which she has done her work has challenged the minds and souls of men.[18]

In this English edition, Marie frames the life story beginning and end and Mrs. Meloney doubly frames the opening by beginning with Marie after Marie's Preface. Thus, Pierre, who never wrote his memoir, is remembered though his wife and someone, like Mrs. Meloney, who is among her admirers. Meloney, as the biography in the finding aid to her papers at Columbia University observes, kept a connection with Marie Curie:

> The French government awarded her the Médaille d'Or for her work in the Better Homes Movement, and, at Marie Curie's request, made her an officer of the Legion of Honor in 1927. The Polish government awarded her the Order of Polonia Restituta for her fundraising work on Marie Curie's behalf.[19]

Before going back to Marie Curie's "Autobiographical Notes," I shall say a little more about Meloney's Introduction as Meloney, a pioneer for women in journalism, was very much interested in the work and contributions of women, which seems to have led her to an interest in Marie Curie.

A few moments in Meloney's Introduction give a sense of the importance of Marie Curie and help to frame Marie Curie's biography of Pierre Curie, distinguished from the French edition I have drawn on here because Marie knew French best in addition to her native Polish. I focus on the English version here because Meloney's support and representation of Marie Curie and her desire that Marie tell her story. Meloney sets the scene in her second paragraph:

> One morning in the spring of 1898, when the United States was going to war with Spain, Madame Curie stepped forth from a crude shack on the outskirts of Paris, with the greatest secret of the century literally in the palm of her hand.
>
> It was one of the silent, unheralded great moments in the world's history.[20]

Here, she represents the dramatic moment of war and discovery of radium, a public and private time. Soon Meloney provides the reason why she is writing the Introduction: "I have been asked to tell why I undertook the Marie Curie Radium Campaign and how I persuaded Madame Curie to write this book."[21] Meloney takes credit for getting Marie to write the book while also undertaking the campaign to raise money to get Marie the radium she needed. Furthermore, Meloney explains her hesitation in writing this prefatory piece, saying of Marie Curie: "'In science,' she said, 'we should be interested in things, not persons.'" Marie, according to Meloney, was reluctant to do life writing:

> Marie Curie is the most modest of women. It is only after long persuasion that she has consented to record the autobiographical notes contained in this book. Still, so much has been left unsaid, uninterpreted, that I feel an obligation to say a word toward a fuller understanding of this great and noble character.[22]

Meloney stresses the modesty of Marie and how long she had to persuade her to write notes of autobiography, while even still, Meloney thinks that more needs to be said and interpreted to understand the greatness and nobility of this "character," Marie. This project was long in the making: "In 1915 I wrote in my editor's suggestion book: 'Greatest woman's story in the world—Marie Curie, discoverer of radium.'"[23] Meloney explains that others sought out Marie Curie but to no avail as she was at the front in the First World War and that when Meloney was working on another book in Paris in 1919, she spoke with Stéphane Lauzanne, who told her to give up because Marie Curie does nothing but work, does not like publicity and focuses on her family, while he also said: "'Her mind is as exact and logical as science itself. She cannot accept or understand exaggerations and inaccurate quotations. She cannot understand why scientists, rather than science, should be discussed in the press.'"[24] Lauzanne also tells of Marie's first lecture as professor at the Sorbonne on 5 October 1906, in which she did not mention her husband but focused on her work, the science of radioactivity, so that, he continued after recounting the opening words of the lecture, that "it is useless to try to interrupt her work for interviews."[25] Apparently, Marie Curie's fellow scientists agreed with Lauzanne, but Meloney persisted, invoking in a letter her own father, a doctor who stressed the importance of people. Meloney, who had met Thomas Edison, a wealthy person, did get to see Marie Curie, in humble circumstances: "I found a simple woman, working in an inadequate laboratory and living in a simple apartment on the meager pay of a French professor."[26] The portrait that Meloney creates is of Marie as a humble, brilliant woman, and she reports a dialogue between them. Meloney wonders whether Marie has patents to have funds for more radium and represents her reply: "'There were no patents. We were working in the interests of science. Radium was not to enrich any one. Radium is an element. It belongs to all.'"[27] Meloney promised to help, and she visited the Institute again and the family apartment on L'île Saint Louis. Meloney tells of her campaign for radium for Marie Curie in the United States and of a telephone conversation in March 1921 between Marie and Lauzanne concerning Marie's prospective trip to the United States. Furthermore, Meloney says:

> A little later, when Madame Curie and I had embarked for America, where she was to receive her radium and other experimental material, I asked her if, the day I had first given her the promise, she had believed that American women would rally to her aid.

Then Meloney states: "'No,' she confessed honestly, 'but I knew you were sincere.'"[28] Meloney shows the selflessness of Marie, how at her marriage she made use of the gift of funds from relatives to buy bicycles for Pierre and her, rather than a fancy dress, and how she had the document changed to make sure the American gift of a gram of radium be given for science and not her family, especially as Marie might die and the radium could be tied up. The agreement was signed on 21 May 1921. Finally, Meloney convinced Marie to write her story, albeit reluctantly. Here is Marie's story in a nutshell as Meloney reports it: "'It is such an uneventful, simple little story. I was born in War-saw of a family of teachers. I married Pierre Curie and had two children. I have done my work in France.'"[29] To end her Introduction, Meloney admits that much has and will be written on the work of Pierre and Marie Curie, but she says that of Marie the woman, the autobiographical notes in this book will be it, and she ends with Marie's view "that 'In science we should be interested in things, not persons.'"[30] This book, then, especially in its English version, is framed by the life of Marie Curie as a biography and autobiography with a life of Pierre Curie. Perhaps this is a creative tension and not simply a contradiction, partly as an aspect of a campaign for a gram of radium to further science and Marie Curie's science in particular.

Here I return after this brief discussion of Meloney's Introduction, which follows the English version of Marie Curie's Preface, which also appears in the French edition, to the "Autobiographical Notes," which is in the English book. In the space I have, I can examine a few strands of Marie's self-representation for which Meloney asked her. Marie is philosophical and thinks about the nature of writing life, including one's own:

> However, I could not conceive my biography as a complete expression of personal feelings or a detailed description of all incidents I would remember. Many of our feelings change with the years, and, when faded away, may seem altogether strange; incidents lose their momentary

> interest and may be remembered as if they have occurred to some other person. But there may be in a life some general direction, some continuous thread, due to a few dominant ideas and a few strong feelings, that explain the life and are characteristic of a human personality. Of my life, which has not been easy on the whole, I have described the general course and the essential features, and I trust that my story gives an understanding of the state of mind in which I have lived and worked.[31]

Marie stresses changing feelings over time, the strangeness of a life as if being remembered as that of another person, the search for a thread to her ideas and feelings as characterizing this human personality while generalizing the course of her life and its essence, comprehending her state of mind in a life that was not easy. Ideas and psychology in a changing person in a changing world represent her focus. Here is a mind reflecting on itself, a person considering her feelings.

She tells the reader that she was born Marie Sklodowska in Poland, about her family of the small and medium landowner class and about her parents, both teachers, and her love of the country and nature. Marie also talks about the death of her sister and her mother: "This catastrophe was the first great sorrow of my life and threw me into a profound depression. My mother had an exceptional personality."[32] Marie then provides a portrait of her exceptional mother who had big heart as well as being intellectual. Marie portrays a family living with this sorrow and the challenges of education after her mother died while also commenting on the Russian control of Poland:

> Warsaw was then under Russian domination, and one of the worst aspects of this control was the oppression exerted on the school and the child. The private schools directed by Poles were closely watched by the police and over-burdened with the necessity of teaching the Russian language even to children so young that they could scarcely speak their native Polish.[33]

She describes this situation: "the moral atmosphere was altogether unbearable."[34] Although Marie says that her father taught physics and mathematics, he read, wrote and translated poetry and read Polish poetry and prose to his family Saturday evenings, which she describes as "a great pleasure and a source of renewed patriotic feelings."[35] Marie brings together education and politics, private and public. Moreover, she represents her own love of poetry, literature and language, national and foreign:

> Since my childhood I have had a strong taste for poetry, and I willingly learned by heart long passages from our great poets, the favorite ones

> being Mickiewecz, Krasinski and Slowacki. This taste was even more developed when I became acquainted with foreign literatures; my early studies included the knowledge of French, German, and Russian, and I soon became familiar with the fine works written in these languages. Later I felt the need of knowing English and succeeded in acquiring the knowledge of that language and its literature.[36]

Marie's self-portrait is one that involves family, culture, nation, science, education, language, literature and poetry. She is proud of Poland and Polish poetry but is also international in culture and literature. Marie says her mother was musical but that she regrets not continuing with music after the death of her mother yet mentions that she found mathematics and physics easy to learn and that her father also helped her yet did not have a laboratory to perform experiments. Trips to the country estate and to Austrian Galicia during vacations allowed an escape from police surveillance. In this part of Austria, "we could speak Polish in all freedom and sing patriotic songs without going to prison."[37] She speaks of her love of plains, mountains, the sea – of nature. She also tells of her brother completing his medical studies and becoming a chief physician at a hospital in Warsaw. Marie and her sister wanted to be teachers like their parents, but her sister became a doctor after studying at the University of Paris. Another sister became a teacher. Marie took a year off from study and at seventeen became a governess with a family on a country estate where she taught and enjoyed nature and the seasons. Marie and the oldest child of the house taught children and little girls in the village who had no access to education, work that "might bring imprisonment or deportation to Siberia."[38] The shadow of Russian oppression is part of Marie's account. She recounts her evening studies and outlines her decision of what path to take:

> I was as much interested in literature and sociology as in science. However, during these years of isolated work, trying little by little to find my real preferences, I finally turned towards mathematics and physics, and resolutely undertook a serious preparation for future work. This work I proposed doing in Paris, and I hoped to save enough money to be able to live and work in that city for some time.[39]

This is a turning point in her life. Talented in humanities, social sciences and science, Marie decides to focus on the subjects she pursued for the rest of her life. For financial reasons, she could not leave with her sister for Paris, and she worked for years in the country and then back in Warsaw. There, in her limited free time, she did experiments in physics and chemistry in a laboratory her cousin directed and learned progress was not easy. Marie also taught evenings as part of a group that would, by the Russian authorities, be considered "a secret organization."[40] Although these limited means could not lead to considerable results, Maric says she still believes that the ideas these teachers shared are the way to social progress:

> You cannot hope to build a better world without improving the individuals. To that end each of us must work for his own improvement, and at the same time share a general responsibility for all humanity, our particular duty being to aid those to whom we think we can be most useful.[41]

Marie stresses individual improvement and responsibility to humanity and duty to be useful and help others. Her science is individual and socially responsible. Marie's father helped her to go live with her sister and her sister's husband in Paris, and her father hoped she would return to live with him in Poland, but her marriage to Pierre meant that fate decided otherwise. Her father, who had wished to do scientific work in his youth, took solace in the success of Marie's work. Instead, her father lived with the family of Marie's brother and helped bring up those grandchildren until his death in 1902.

In 1891, at twenty-four, Marie moved to Paris and was soon in a room near the schools where she lived as a student for four years. Her description includes this assessment:

> I was entirely absorbed in the joy of learning and understanding. Yet, all. the while, my living conditions were far from easy, my own funds being small and my family not having the means to aid me as they would have liked to do. However, my situation was not exceptional; it was the familiar experience of many of the Polish students whom I knew.[42]

Marie balances the joy and the challenges, the oxymoronic painful charm, and adds: "If sometimes I felt lonesome, my usual state of mind was one of calm and great moral satisfaction."[43] At the Sorbonne, Marie also notes that she had to catch up with French students in physical sciences and mathematics. She recounts the delight of discovery: "All that I saw and learned that was new delighted me. It was like a new world opened to me, the world of science, which I was at last permitted to know in all liberty."[44] She lived a basic life to be free to study science and describes her fellow students and how well she did on the examinations in 1893 and 1894 while saying that this period in her life is one of her best memories. Then she tells of meeting Pierre Curie in 1894. The scene is similar to the one she described in her biography of Pierre. She also goes over similar ground about Pierre and his work and family, how Pierre and Marie got along so well and how close he was to his parents and older brother and how she "had the privilege of entering into a family worthy of affection and esteem, and where I found the warmest welcome," and they had a simple wedding and found a quiet apartment.[45] The solitary life ends.

Chapter 2 stresses the end of solitude and the beginning of a new life. Marie's representation of her life with Pierre after the marriage resembles that in the biography of Pierre. Family is a key part to Marie's view of Pierre and herself in the biography and autobiography. She talks here as she did in the biography about her work in the laboratory, her studies, how Pierre and she would stay with his parents, their love of the country, their vacations. Once more, she mentions Pierre learning some Polish on their vacation in the Carpathian Mountains. She records here her scientific work and publications, the birth of their eldest daughter, the death of Pierre's mother, that Pierre's father came to live with them in their house with a garden, their quiet life. Marie also tells of her doctoral work inspired by the work of Henri Becquerel, which she explains in some detail, mentioning that Pierre was also excited by Becquerel's findings. Moreover, Marie details the Curies' unexpected discoveries: "Neither of us could fore-see that in beginning this work we were to enter the path of a new science which we should follow for all our future."[46] Not only is her account illuminating for their work but also for how science works: "Here is, briefly, the story of the search and discovery."[47] Part of this clear and helpful description is a reminder, as she had done in the biography, of the struggle, the lack of resources: "We had no money, no suitable laboratory, no personal help for our great and difficult under-taking."[48] Despite the obstacles, Marie remembers these days fondly: "Yet it was in this miserable old shed that we passed the best and happiest years of our life, devoting our entire days to our work."[49] As in the biography, in the autobiography, the story is a touching one – the two accounts overlap. Pierre and Marie worked together but sometimes with different emphasis: "At this stage I devoted myself especially to the purification of the radium, my husband being absorbed by the study of the physical properties of the rays emitted by the new substances."[50]

She also recounts other details that can be found in the biography, for instance, how they did not go to Geneva but stayed in Paris and how Marie directed the studies of young women in physics. Marie notes that their quiet life ended because of the announcement of the discoveries of the Curies. In 1903, Marie finished her doctorate at the Sorbonne and was awarded with Becquerel and Pierre, the Nobel Prize "for the discovery of radioactivity and new radioactive elements."[51] The Prize, an honour, disturbed their peace but provided funds for their research, and Marie also tells of their exhaustion and failing health, which prevented them going to Stockholm until 1905. Marie stresses the "real suffering" caused by these external disruptions of their "voluntary isolation" – which she sees as freedom – while also telling of the birth of their second daughter, Eve Denise, in 1904 and that year Pierre became professor of Physics at the Sorbonne and Marie chief of his laboratory.[52] She also tells of "the dreadful catastrophe" of Pierre's death and how she was left with the children to continue the scientific work Pierre and she had done.[53]

One of the aspects of nature, literature and culture I am stressing here and in the book as a whole, including discussions of scientists and their families, is feeling. Whether natural selection, botany, radium and radioactivity, logic or algorithms, scientists, like poets, artists, critics and humanists, think but feel at the same time. This is something obvious that can be forgotten, especially when discussing theory, method and machine learning or artificial intelligence. Nature and human nature are intertwined in human culture. Humans are of and from nature, conscious of nature, both being at one with it and being estranged from it. Nature bears us and sheds us.

Marie Curie is not afraid, as logical and as committed to science as she is, to show her feelings:

> It is impossible for me to express the profoundness and importance of the crisis brought into my life by the loss of the one who had been my closest companion and best friend. Crushed by the blow, I did not feel able to face the future. I could not forget, however, what my husband used sometimes to say, that, even deprived of him, I ought to continue my work.[54]

Marie uses words to express the expression that words cannot express – the topos of inexpressibility – to represent the crisis coming from Pierre's death. She describes this death also as a "national misfortune" and observes:

> It was largely under the influence of this emotion that the Faculty of Sciences of Paris decided to offer me the chair, as professor, which my husband had occupied only one year and a half in the Sorbonne. It was an exceptional decision, as up to then no woman had held such a position.[55]

She stresses the role of private and public feelings – grief – as a way of showing "esteem" and a "gift" for her to continue her research, allowing a woman to become a professor of physics at the Sorbonne for the first time.[56] Marie often stresses being a girl and then a woman, a daughter, a sister, a wife and a mother. She is forthright about difficulties and challenges:

> I had not expected a gift of this kind; I never had any other ambition than to be able to work freely for science. The honor that now came

> to me was deeply painful under the cruel circumstances of its coming. Besides I wondered whether I would be able to face such a grave responsibility.[57]

Marie was devoted to science, and it is understandable that like anyone with devotion to science, art or scholarship, she wished to focus on that work. She tells of "much hesitation" she experienced before beginning her teaching at the Sorbonne in 1906 and of how, moving to a house with a garden in Sceaux, a suburb of Paris, she had to raise her children with the help of Pierre's father, who shared with her an "inner grief."[58] One of Marie's cousins and then a woman who helped to bring up the daughter of one of Marie's sisters helped out. Marie's daughter learned Polish and her family came to visit and to go on vacations with Marie and her family. Moreover, Marie speaks about how deeply the death of her father-in-law in 1910 after a long illness affected Marie and her elder daughter, who was twelve. Marie discusses the education of her elder and younger daughters, the one given to science and having a mind like Pierre's and the younger had a gift for music. In this account, Marie includes her views of education, saying that lycées held long class hours that were detrimental to health and says:

> My view is that in the education of children the requirement of their growth and physical evolution should be respected, and that some time should be left for their artistic culture. In most schools, as they exist to-day, the time spent in various reading and writing exercises is too great, and the study required to be done at home too much. I also find these schools lacking, in general, in practical exercises to accompany the scientific studies.[59]

Marie stresses nature, exercise, free time at home and experiment or practical exercises. With like-minded friends, for two years, Marie formed a cooperative in which they all helped to teach their children, and she says that the children were interested in the practical scientific exercises. Marie notes that the student benefitted from this schooling, her elder daughter being advanced and later studying science at the Sorbonne. The younger daughter did well but did not benefit from this cooperative. Marie talks about how important physical education for girls is and how she made sure that her daughters did gymnastics and sports and took vacations by the sea and in the mountains, swimming and canoeing, walking and bicycling. She addresses the balance between work and family life:

> But of course the care of my children's education was only a part of my duties, my professional occupations taking most of my time. I have been

> frequently questioned, especially by women, how I could reconcile family life with a scientific career. Well, it has not been easy; it required a great deal of decision and of self-sacrifice. However, the family bond has been preserved between me and my now grown-up daughters, and life is made brighter by the mutual affection and understanding in our home, where I could not suffer a harsh word or selfish behavior.[60]

Marie never says any of this is easy, and she focused on her career but also values the family bond. This goal of mutual affection and understanding is something she stresses and has thought about. She shifts back to her working conditions at the Sorbonne in 1906, her provisional laboratory, but had help from scientists and students to continue her research. She praises Andrew Carnegie who donated significantly to Marie's laboratory and talks about her preparation of radium chloride, her attempts to isolate polonium, her methods of measurement of radium, "radiumtherapy (in France, *Curietherapy*)."[61] Part of Marie's narrative is the benefit of the research that the Curies and their associates did:

> It may be easily understood how deeply I appreciated the privilege of realizing that our discovery had become a benefit to mankind, not only through its great scientific importance, but also by its power of efficient action against human suffering and terrible disease. This was indeed a splendid reward for our years of hard toil.[62]

She stresses her deep appreciation of the benefits the research has for humankind and devotes space to measurement standards.

Then she shifts to honours. In 1910, she refused the Legion of Honour as did her husband, who refused honorary distinctions. As he had become a member of the Academy of Sciences in the last months of his life, she allowed herself to be a candidate, a first as a woman, but she fell short in votes, and she says she will not renew her candidacy. She does not like elections that involve "personal solicitation" and has been elected to other academies that do not require that soliciting.[63] Ironically, she received a second prize:

> As a result of all the cares devolving on me, I fell seriously ill at the end of 1911, when, for the second time, I received, this time alone, the award of the Nobel prize. This was a very exceptional honor, a high recognition of the discovery of the new elements and of the preparation of pure radium. Suffering though I was, I went to Stockholm to receive the prize.[64]

Her elder sister and elder daughter, Irene, accompanied her, and women in Sweden held a reception for her. Back in France, she was ill and in bed for months, then having to move from Sceaux to Paris for the education of her daughters. In 1912, she collaborated in creating a laboratory of radium in Warsaw but could not leave France to be the director, but in 1913, she was able to go to the inauguration, which she remembers fondly despite the difficult political situation there then. When she recovered her health, she also renewed her efforts to have a laboratory built in Paris, which was not completed from 1912 to the outbreak of war in 1914, something she chronicles.

In Chapter 3, Marie Curie discusses the war as it affected the nation, her family and her. She notes:

> The historic events that followed are known to every one, but only those who lived in Paris through the days of August and September, 1914, can ever really know the state of mind in the capital and the quiet courage shown by it.[65]

She explains how the government moved to Bordeaux and how she was told to take her radium there. Marie describes her work in the war, including this passage:

> The work was the organization of radiologic and radio-therapeutic services for the military hospitals. But I also had to make the change, during these difficult war years, of my laboratory into the new building of the Institute of Radium and to continue, in the measure possible to me, regular teaching, as well as to investigate certain problems especially interesting the military service.[66]

Marie is always working hard, and she discusses the application of X-rays in war and peace. For instance, she speaks of the work that her daughter, Irene, and she did: "But nothing was so moving as to be with the wounded and to take care of them. We were drawn to them because of their suffering and be-\cause of the patience with which they bore it."[67] This suffering she addresses and describes in more detail:

> I can never forget the terrible impression of all that destruction of human life and health. To hate the very idea of war, it ought to be sufficient to see once what I have seen so many times, all through those years: men and boys brought to the advanced ambulance in a mixture of mud and blood, many of them dying of their injuries, many others recovering but slowly through months of pain and suffering.[68]

The shock of how war destroys health and life, she stresses and shares with the reader as her abhorrence of war, which she hated but whose destruction and suffering she tried to alleviate. She also wrote about radiology in war and peace:

> My continued and various experience in war radiology gave me a wide knowledge of that subject, which I felt should be made more familiar to the public. So I wrote a small book called 'Radiology and the War,' in which I aimed to demonstrate the vital importance of radiology and to compare its development during war time with its use in the previous time of peace.[69]

Marie is a researcher but also a writer and a teacher of science. She teaches her students and the public.

Marie Curie talks about her laboratory and shows how important plants and nature are to her:

> One of my first cares was to have trees planted in the limited grounds of my laboratory. I feel it very necessary for the eyes to have the comfort of fresh leaves in spring and summer time. So I tried to make things pleasant for those who were to work in the new building. We planted a few lime trees and plane trees, as many as there was room for, and did not forget flowerbeds and roses.[70]

Trees and nature generally were restorative for Marie as they were for Pierre. She has a sense of health for body and spirit. Even after the war, there were challenges:

> After more than four years of a war which caused ravages without precedent, the armistice came at last, in the autumn of 1918, followed by laborious efforts to reëstablish peace, which is not yet general nor complete. It was a great relief to France to see the end.[71]

She does celebrate the establishment of free and independent Poland and speaks of her visit there: "In these new conditions I went to Warsaw and saw my family again, after many years of separation, in the capital of free Poland."[72] Marie also says that the war devastated France and had affected the laboratories and the Radium Institute. She wonders about the future:

> I myself am no longer young, and I frequently ask myself whether, in spite of recent efforts of the government aided by some private donations, I shall ever succeed in building up for those who will come after me an Institute of Radium, such as I wish to the memory of Pierre Curie and to the highest interest of humanity.[73]

Despite her doubts, she continues and tells of how Mrs. Meloney helped to create the Marie Curie Radium Fund and invited Marie and her daughters to the United States and to accept the gram of radium from President Harding at the White House. Marie speaks of the "grateful memory of my sojourn."[74] She did many things in the United States and visited universities, women's colleges and groups, Niagara Falls, the Grand Canyon, saying:

> Unhappily, the precarious state of my health did not permit of the complete fulfilment of the general plan established by my visit to America. However, I saw and learned much, and my daughters enjoyed to a full extent the opportunities of their unexpected vacation and the pride in the recognition of their mother's work.[75]

She states that she faced challenges on her return to France and thought about how a scientist should consider his discovery. Marie also notes how Pierre and she viewed their discoveries:

> My husband, as well as myself, always refused to draw from our discovery any material profit. We have published, since the beginning, without any reserve, the process that we used to prepare the radium. We took out no patent and we did not reserve any advantage in any industrial exploitation. No detail was kept secret, and it is due to the information we gave in our publications that the industry of radium has been rapidly developed. Up to the present time this industry hardly uses any methods except those established by us. The treatment of the minerals and the fractional crystallizations are still performed in the same way, as I did it in my laboratory, even if the material means are increased.[76]

The Curies did not patent and profit from their discoveries but shared them. Theirs is a view of science as knowledge and not a matter of property and profit. For Marie, humanity needs scientists who are dreamers not focused on material benefit, but that dreamers need a life free of "material care" to devote themselves "to the service of scientific research."[77] She balances two sides of the public good.

Chapter 4 is about the trip to the United States, and Marie Curie revisits this in more detail. She mentions Mrs. Meloney again. More generally, Marie says: "America not only gave me a generous welcome, but also true friends whom I could not thank enough for their kindness and their devotion."[78] She describes the women's colleges, her visit to New York and the White House at some length. For instance, of this reception in Washington, DC, Marie says: "A remembrance never to be forgotten was left by this reception in which the chief representative of a great nation offered me homage of infinite value, the testimonial of the recognition of his country's citizens."[79] She praises the United States for recognizing something for the general interest, this spirit that animates science in France. Women and gender also concern Marie. Part of Marie's account is about how American men encourage American women in their aspirations. She is cautious about generalizations not based on adequate observation but sometimes offers some observations:

> To make complete these travel impressions it would be necessary to speak of the nature of the country. I recoil before the task, being incapable of expressing in a few words the immensity and the variety of the spaces which opened before my eyes. The general impression is one of unlimited possibilities for the future.[80]

This typology between France and the United States is something found between places in Herodotus, Tacitus, Montaigne and others. Comparison helps in an understanding about science in the United States and France. Marie represents the departure for France and reiterates her hesitation to generalize:

> I would not take the liberty, after so short a period of time, of giving an opinion on America and the Americans. I would only say how deeply I have been touched by the warm reception which was tendered everywhere to me and my daughters.[81]

She represents the friendship with those she met in the United States and the French between the two countries, "a feeling of affection" and "a mutual sympathy which gives confidence in a peaceful future for humanity."[82] Peace is part of her vision of science, although she has experienced the devastation of war up close. And thus ends the "Autobiographical Notes" that ends the English edition.

Among her writings, Marie Curie also wrote a preface to the volume that collected the work of Pierre Curie. As can be seen in her manuscripts, Marie also wrote notes on her daughters when they were young.[83] Marie's handwritten manuscript notes begin thus: "Irène Curie née le 12. Septembre 1897. à 10 heures du Soir" ("*Irène Curie* born on 12. September 1897 at 10 o'clock in the evening"), and on the next pages her mother measures from 1897 to 1913 her height in centimetres, reaching 166 cm. that year, but also recording her weight at birth as 3 kg and weighing her until 1913.[84] Marie notes a good deal in numbers and observes her daughter, for instance when she begins to use her hand with objects and other milestones in development: "Elle a commencé à avoir peur des personnes et des objets inconnus, des grosses voix, etc." ("She started to be afraid of unknown people and objects, loud voices, etc.").[85]

Marie's manuscript discusses different matters, including *Irène's study of mathematics and languages.* One of the aspects that I have not mentioned and was not part of the texts that Marie wrote about Pierre and herself or the biography that Eve wrote about her was "l'affaire Langevin" in 1911.[86] Marie reports *Irène's response to some of the French press alleging an affair between Marie and Paul Langevin, Pierre's former student, and meant that there was pressure on the Curie family and name.* In these manuscript notes about Irène, Marie mentions this matter and Irène's discovery of it:

> Elle est très s'emotionée par la publication du Journal sur 'l'affaire Langevin. Elle a vu cet odieux journal en allant au Collège et y est restée sans rien dire à personne, mais absorbée absente. Puis elle est sortie et a pleuré. Elle a ensuite été chez les Perrin et avec Mme Perrin elle est allée voir André Debierne. Ses amis l'ont console peu à peu. Mais elle a été très touché. (She was very upset by the publication of the Journal on the 'Langevin Affair'. She saw this odious newspaper on her way to the College and stayed there without saying anything to anyone, but absorbed as if absent. Then she went out and cried. She then went to the Perrins' and with Mrs. Perrin she went to see André Debierne. Her friends consoled her little by little. But she was very touched).[87]

Irène, who had just entered her teens, discovered this scandal on the way to school. After another paragraph, Marie mentions someone who seems to have been so important to Pierre and Marie and now to her elder daughter (and, as we shall see, also for Eve): "A Pâques je l'ai envoyée à Montpellier, a cause du grand désir qu'elle a de voir son oncle" ("At Easter I sent her to Montpellier, because of the great desire she has to see her uncle").[88] Marie also notes this about Irène: "elle adore tout ce qui est relative à la famille Curie" ("she loves everything related to the Curie family"), and Marie stresses "la famille de son père" ("the family of her father") while saying; "A Montpellier elle est très heureuse" ("At Montpellier she is very happy") and that she then sees the Borels with Maurice Curie, her cousin and the son of Jacques Curie who two years later worked with Marie in her laboratory and who was later a professor of physics at the Sorbonne.[89] The Perrins and Debierne were part of the scientific circle of friends of the Curies. Jean Perrin won the Nobel Prize for Physics in 1926. There are many twists and turns in this story, and I am focusing most on science, but science and scientists also have to face ignorance, prejudice and gossip and scientists, like all of us, are all too human. Nanny Fröman notes that Marguerite Borel's book quotes the words of Jean Perrin:

> 'But for the five of us who stood up for Marie Curie against a whole world when a landslide of filth engulfed her, Marie would have returned

> to Poland and we would have been marked by eternal shame.' The five were Jean and Henriette Perrin, Émile and Marguerite Borel and André Debierne

and Fröman says that there were never any legal proceedings and that Langevin and his wife settled without mentioning Marie's name:

> We shall never know with any certainty what was the nature of the relationship between Marie Curie and Paul Langevin. It is referred to by Paul Langevin's son, André Langevin, in his biography of his father, which was published in 1971. He writes, 'Is it not rather natural that friendship and mutual admiration several years after Pierre's death could develop step by step into a passion and a relationship?' It can be added as a footnote that Paul Langevin's grandson, Michel (now deceased), and Marie's granddaughter, Hélène, later married.[90]

There are many footnotes and future threads after the deaths of Pierre in 1906 and Marie in 1934. Family members of what I might call the Curie circle were still interpreting these matters from generations before and were intertwined. Marie's daughter, Irène, won the Nobel Prize, with her husband, in 1935, the year after her mother died. The Nobel biography of her notes that she married Frédéric Joliot in 1926, the year after she became Doctor of Science with a thesis on the alpha rays of polonium, saying that alone with her husband, she did significant

> work on natural and artificial radioactivity, transmutation of elements, and nuclear physics; she shared the Nobel Prize in Chemistry for 1935 with him, in recognition of their synthesis of new radioactive elements, which work has been summarized in their joint paper *Production artificielle d'éléments radioactifs. Preuve chimique de la transmutation des éléments* (1934).[91]

Marie Curie also has left manuscript notes on Eve, her younger daughter, which begins "Eve Denise née le 6 Decembre 1904 à 9 heures du matin" ("Eve Denise born the 6 of December 1904 at the 9th hour of the morning").[92] Marie measures her weight (from at birth 3, 730 [kg.] to that on 21 September 1915 at 37.5) and height (82.1/2 cm on 1 October 1906 to 132 cm on 26 July 1913).[93] On 30 October 1905, for instance, Marie records the following about Eve: "Elle fait ses premiers pas, elle fait à plusieurs reprises un trajet de 3 petits pas" ("She takes her first steps, she takes 3 small steps several times").[94] As part of an entry for the summer of 1908, Marie observes about Eve: "Elle a des dispositions remarquables pour la musique" ("She had remarkable dispositions for music" and ends the entry with "Commence à lire et à écrire" ("Begins to read and to write").[95] Marie includes in her entry on Eve for the spring of 1911 a further assessment: "Elle lit et écrit bien" ("She reads and writes well"); then she adds something about her character: "Elle est très bonne, défend tout le monde, ne peut voir pleurer ni souffrir, désire faire plaisir aux autres. Coeur exquis et grande sensibilité" ("She is very good, defends everyone, cannot see crying or suffering, desires to please others. Exquisite heart and great sensitivity").[96] In the entry of 9 June 1911, Marie speaks about Eve's musical talent and writes in French and Polish.[97] After writing about Eve's vacation in England in August and September 1912, Marie discusses her return and ends her manuscript on her younger daughter with what seems to be an allusion to Eve being with her uncle Jacques (she gives the first name but does not mention uncle or Curie, so I am making an assumption) and her walking and bicycling ("elle se promène et fait de la bicyclette").[98] As with Irène, with Eve, Marie ends the narrative in 1912 and leaves blank pages.

Eve wrote a biography of her mother, published in 1937. Some of this work I hope to analyse in the future, but I end here with quotations from the English translation of Eve's biography, which may be more available to the reader of my book. Eve begins her Introduction thus: "THE LIFE OF MARIE CURIE contains prodigies in such number that one would like to tell her story like a legend," adding that "It would have been a crime to add the slightest ornament to this story, so like a myth," while Eve ends her Introduction with her words and the words from a colleague and friend of her mother:

> I should have liked the gifts of a writer to tell of this eternal student—of whom Einstein said: 'Marie Curie is, of all celebrated beings, the only one whom fame has not corrupted'—passing like a stranger across her own life, intact, natural and very nearly unaware of her astonishing destiny.[99]

Although Eve uses "legend" and "myth" in connection with telling her mother's story, she does not wish to embellish the story with additions. Like Eve, I quote Einstein, here from the memorial in New York in 1935 for Marie Curie, quoted in an article of 1999 by Stanley C. Pycior, who notes – "(The author wishes to thank Madame Eve Curie Labouisse, who previewed the article before submission for publication)"[100]

> It was my good fortune to be linked with Mme. Curie through twenty years of sublime and unclouded friendship. I came to admire her human grandeur to an ever-growing degree. Her strength, her purity of will, her

> austerity toward herself, her objectivity, her incorruptible judgment – all these were of a kind seldom found joined in a single individual. She felt herself at every moment to be a servant of society, and her profound modesty never left any room for complacency.[101]

Born in 1904 in Paris, Eve Curie died in New York in 2007. Born in 1897 in Paris, Irène Curie died there in 1956, a year after Einstein, who knew Marie and her daughters. Pierre had died in 1906 and never met Einstein. It was not so long ago that Eve died, and the story of the Curies continues.

When in Paris, I live around the corner from the Curie Museum, which describes itself thus:

> Le Musée Curie se situe au 1, rue Pierre et Marie Curie, dans le 5e arrondissement de Paris au cœur du « Campus Curie » dans le quartier de la Montagne Sainte Geneviève, à deux pas du Panthéon où reposent Pierre et Marie Curie depuis 1995. (The Curie Museum is situated at 1, rue Pierre et Marie Curie, in the 5th arrondissement of Paris in the heart of the "Campus Curie" in the Montagne Sainte Geneviève district, a stone's throw [a few steps] from the Pantheon where Pierre and Marie Curie have been buried since 1995).[102]

Over the years, I have been in that fine museum many times and had the pleasure of teaching a doctoral seminar in the old Sorbonne building. My grandfather went to school not far away. The Curies are at the heart of this remarkable part of this remarkable city with its gardens and its cultural and scientific heritage. Paris has inspired many over the generations and has prompted me to write in French and English in poetry and prose. Ideas, as Marie Curie knew, are worth pursuing, and reluctantly she might admit, as she did not always do so directly, that her kindness and devotion to her family and friends was also an important part of her gift to humanity.

Notes

1 My translation; Marie Curie, *Pierre Curie* (Les grands hommes de France, collection dirigée par Paul Gautier) (Paris: Payot, 1924), 98–99; see Marie Curie, *Pierre Curie*, translated by Charlotte Kellogg and Vernon L. Kellogg, With An Introduction by Mrs. William Brown Meloney and Autobiographical Notes by Marie Curie (New York: The Macmillan Company, 1923), 137.

2 On the death of Pierre Curie, see "MORT TRAGIQUE DU DR CURIE ÉCRASÉ PAR UN CAMION," *L'Humanité : Journal Socialiste Quotidien*, Friday 5 April 1906, front page; electronic version at https://gallica.bnf.fr/ark:/12148/bpt6k2509147/f1.zoom.langFR#; see also "The Paris Archives: The Death of Pierre Curie," *Invisible Paris*, https://parisisinvisible.blogspot.com/2012/07/paris-archives-death-of-pierre-curie.html ; see also Marie-Louise Concasty, *Pierre et Marie Curie*: [exposition] (Paris, Bibliothèque nationale,

[octobredécembre] 1967); Karen Hays, "Auto-Duet: Essays on Competence and Acoustics," *The Georgia Review* 68 (4) (2014): 696–697.

3 *Pierre Curie* (Les grands ...), 99; my translation; see Curie, *Pierre Curie*, 137–138.

4 Eve Curie, *Madame Curie: A Biography by Eve Curie*, translated by Vincent Sheean (Garden City: Doubleday, Doran & Company, Inc, 1938), 246.

5 Curie, *Madame Curie*, 246.

6 Curie, *Madame Curie*, 247.

7 *Pierre Curie* (Les grands ...), 102; my translation; see Curie, *Pierre Curie*, 142.

8 *Pierre Curie* (Les grands ...), 103; my translation; see Curie, *Pierre Curie*, 143.

9 *Pierre Curie* (Les grands ...), 103–104; my translation; see Curie, *Pierre Curie*, 143–144.

10 *Pierre Curie* (Les grands ...), 104–105; my translation; see Curie, *Pierre Curie*, 143.

11 *Pierre Curie* (Les grands ...), 106; my translation; see Curie, *Pierre Curie*, 146.

12 *Pierre Curie* (Les grands ...), 108; my translation; see Curie, *Pierre Curie*, 147.

13 My translation; *Pierre Curie* (Les grands ...), 108–109; see Curie, *Pierre Curie*, 148–149.

14 My translation; *Pierre Curie* (Les grands ...), 110; see Curie, *Pierre Curie*, 150.

15 *Pierre Curie* (Les grands ...), 111; my translation; see Curie, *Pierre Curie*, 151.

16 For "repousser" in the electronic version of Larousse, see https://www.larousse.fr/dictionnaires/francais/repousser/68469#:~:text=repousser%20sa%20barbe.-,repousser%20v.t.,Pousser%2C%20cro%C3%AEtre%20de%20nouveau.

17 Marie Curie, "Autobiographical Notes," in *Pierre Curie*, translated by Charlotte Kellogg and Vernon L. Kellogg, 155.

18 Mrs. William Brown Meloney, "The Life Story of Pierre Curie: Introduction," in Marie Curie, *Pierre Curie*, translated by Charlotte Kellogg and Vernon L. Kellogg, With an Introduction by Mrs. William Brown Meloney and Autobiographical Notes by Marie Curie (New York: The Macmillan Company, 1923), 11. See the article on the life of Marie Meloney at her death: "MRS. William Brown Meloney," *The New York Times*, 25 June 1943, page 16. On Meloney, also see "The Press: Herald Tribune's Lady," TIME, October 8, 1934, https://time.com/archive/6767460/the-press-herald-tribunes-lady/.

19 Marie Mattingly Meloney papers 4079096, page 5, Marie Mattingly Meloney papers, 1891–1943, Archival Collections, Columbia University Libraries, https://findingaids.library.columbia.edu/archives/cul-4079096.

20 Meloney, "The Life," 11.

21 Meloney, "The Life," 12.

22 Meloney, "The Life," 12.

23 Meloney, "The Life," 12.

24 Meloney, "The Life," 13.

25 Meloney, "The Life," 15.

26 Meloney, "The Life," 16.

27 Meloney, "The Life," 18.

28 Meloney, "The Life," 22.

29 Meloney, "The Life," 26.

30 Meloney, "The Life," 30.

31 Curie, "Autobiographical Notes," 155.

32 Curie, "Autobiographical Notes," 157.

33 Curie, "Autobiographical Notes," 158.

34 Curie, "Autobiographical Notes," 159.

35 Curie, "Autobiographical Notes," 160.
36 Curie, "Autobiographical Notes," 160.
37 Curie, "Autobiographical Notes," 161.
38 Curie, "Autobiographical Notes," 165.
39 Curie, "Autobiographical Notes," 165–166.
40 Curie, "Autobiographical Notes," 168.
41 Curie, "Autobiographical Notes," 168.
42 Curie, "Autobiographical Notes," 170.
43 Curie, "Autobiographical Notes," 171.
44 Curie, "Autobiographical Notes," 171.
45 Curie, "Autobiographical Notes," 175.
46 Curie, "Autobiographical Notes," 182.
47 Curie, "Autobiographical Notes," 183.
48 Curie, "Autobiographical Notes," 185.
49 Curie, "Autobiographical Notes," 186.
50 Curie, "Autobiographical Notes," 188.
51 Curie, "Autobiographical Notes," 190.
52 Curie, "Autobiographical Notes," 190, see 191.
53 Curie, "Autobiographical Notes," 191.
54 Curie, "Autobiographical Notes," 191–192.
55 Curie, "Autobiographical Notes," 192.
56 Curie, "Autobiographical Notes," 192.
57 Curie, "Autobiographical Notes," 192.
58 Curie, "Autobiographical Notes," 192–193.
59 Curie, "Autobiographical Notes," 194–195.
60 Curie, "Autobiographical Notes," 196–197.
61 Curie, "Autobiographical Notes," 199.
62 Curie, "Autobiographical Notes," 200.
63 Curie, "Autobiographical Notes," 202–203.
64 Curie, "Autobiographical Notes," 203.
65 Curie, "Autobiographical Notes," 206.
66 Curie, "Autobiographical Notes," 209.
67 Curie, "Autobiographical Notes," 215.
68 Curie, "Autobiographical Notes," 216.
69 Curie, "Autobiographical Notes," 218.
70 Curie, "Autobiographical Notes," 220.
71 Curie, "Autobiographical Notes," 221.
72 Curie, "Autobiographical Notes," 222.
73 Curie, "Autobiographical Notes," 223–224.
74 Curie, "Autobiographical Notes," 224.
75 Curie, "Autobiographical Notes," 225.
76 Curie, "Autobiographical Notes," 226.
77 Curie, "Autobiographical Notes," 227.
78 Curie, "Autobiographical Notes," 230.
79 Curie, "Autobiographical Notes," 234.
80 Curie, "Autobiographical Notes," 241.
81 Curie, "Autobiographical Notes," 241–242.
82 Curie, "Autobiographical Notes," 242.
83 Pierre Curie, *OEUVRES DE PIERRE CURIE: Publiees par les soins de la Societe Francaise de Physique* (Paris: Guthier-Villars, 1908); Curie, Pierre (1859–1906). Producteur d'un fonds d'archives. Pierre et Marie Curie.

Papiers. V – DOCUMENTS A CARACTERE PRIVE. CLII Marie Curie. Notes sur l'enfance de ses filles. 1897–1912. 1801–2000. Gallica, BNF.

84 Marie Curie, "Notes sur l'enfance de ses filles, 1897–1912, 20 pages sure Irène," page 1, see pages 2–4.

85 Marie Curie, "Notes … Irène," 7.

86 On this relation with Langevin, see Karin Blanc, « Le couple Curie et les prix Nobel », *Bibnum* [En ligne], Physique, mis en ligne le 15 juillet 2018, consulté le 23 octobre 2024. URL: http://journals.openedition.org/bibnum/1172; DOI: https://doi.org/10.4000/bibnum.1172; Eva Hemmungs Wirtén, "Scandal, Slander, and Science: Surviving 1911," in *Making Marie Curie: Intellectual Property and Celebrity Culture in an Age of Information* (Chicago: University of Chicago Press, 2015), 41–74. Chicago Scholarship Online, 2015. https://doi.org/10.7208/chicago/9780226235981.003.0002.

87 Marie Curie, "Notes sur l'enfance de ses filles," 1897–1912, 20 pages sure Irène, pages 16–17.

88 Marie Curie, "Notes sur l'enfance de ses filles," 1897–1912, 20 pages sure Irène, page 17.

89 Marie Curie, "Notes sur l'enfance de ses filles," 1897–1912, 20 pages sure Irène, page 17.

90 Nanny Fröman, "Marie and Pierre Curie and the discovery of polonium and radium," The Nobel Prize, "Originally delivered as a lecture at the Royal Swedish Academy of Sciences in Stockholm, Sweden, on February 28, 1996. Translation from Swedish to English by Nancy Marshall-Lundén," https://www.nobelprize.org/prizes/themes/marie-and-pierre-curie-and-the-discovery-of-polonium-and-radium/.

91 See the Nobel Prize website at https://www.nobelprize.org/prizes/chemistry/1935/joliot-curie/biographical/.

92 Marie Curie. Notes sur l'enfance de ses filles, 1897–1912, 12 pages sur Eve – 1904 à 1912, 1801–2000, Gallica, BNF, page 29, see pages 30–37.

93 Marie Curie, Notes sur l'enfance de ses filles, 1897–1912, 12 pages sur Eve, 30–31.

94 Marie Curie, Notes sur l'enfance de ses filles, 1897–1912, 12 pages sur Eve, 33.

95 Marie Curie, Notes sur l'enfance de ses filles, 1897–1912, 12 pages sur Eve, 34.

96 Marie Curie, Notes sur l'enfance de ses filles, 1897–1912, 12 pages sur Eve, 35.

97 Marie Curie, Notes sur l'enfance de ses filles, 1897–1912, 12 pages sur Eve, 35–36.

98 Marie Curie, Notes sur l'enfance de ses filles, 1897–1912, 12 pages sur Eve, 37.

99 Curie, *Madame Curie*, v and xi.

100 Stanley W. Pycior, "Marie Skłodowska Curie and Albert Einstein: A Professional and Personal Relationship," *The Polish Review* 44 (2) (1999): 142, see 131–142.

101 Pycior, "Marie Skłodowska Curie and Albert Einstein," 141.

102 See the museum website at https://musee.curie.fr/; my translation.

9 Bertrand Russell

Religion, Mind and Autobiography

An analysis of a few key works follows. Portraits and self-portraits were part of the story of the Darwins, Grays, Thaxters and Curies. Part of my emphasis in my book, and in this chapter, is how figures present themselves or are presented. Russell had autobiographical writings from which I will select to discuss. In *My Religious Reminiscences*, Russell speaks of his parents and his grandparents in relation to their ideas and his grandmother, once a Scotch Presbyterian, raising him as a Unitarian.[1] Russell observes:

> when I was about fourteen my thoughts turned to theology. During the four following years I rejected, successively, free will, immortality, and belief in God, and believed that I suffered much pain in the process, though when it was completed I found myself far happier than I had been while I remained in doubt.[2]

Religion and theology mattered to Russell whatever his beliefs. Part of those considerations had to do with science and evolution. Despite some of the restrictions owing to religion, the household he grew up had some latitude:

> Nevertheless on most topics the atmosphere was liberal. For instance, Darwinism was accepted as a matter of course. I had at one time, when I was thirteen, a very orthodox Swiss tutor, who, in consequence of something I had said, stated with great earnestness: 'If you are a Darwinian I pity you, for one cannot be a Darwinian and a Christian at the same time.' I did not then believe in the incompatibility, but I was already clear that, if I had to choose, I would choose Darwin.[3]

Evolution and Christianity were not, for the young Russell, incompatible. Charles Darwin comes up in Russell's thought at an early age, but Russell tries to give a sense of his own views as an undergraduate:

DOI: 10.4324/9781003651871-10

> Until I went to Cambridge I was almost wholly unaware of contemporary movements of thought. I was influenced by Darwin, and then by John Stuart Mill, but more than either by the study of dynamics; my outlook, in fact, was more appropriate to a seventeenth- or eighteenth-century Cartesian than to a post-Darwinian.[4]

Russell says that at Cambridge he learned about the modern world, with McTaggart making him a Hegelian and G. E. Moore having Russell revert to the views he had before Cambridge when he read in his family library.[5] Russell discusses Hegel in relation to God and to truth and falsehood, which for Hegel is not absolute, and the Absolute Idea swallows God.[6] According to Russell, Moore "found the Hegelian philosophy inapplicable to chairs and tables, and I found it inapplicable to mathematics; so with his help I climbed out of it, and back to common sense tempered by mathematical logic."[7] Russell came to the conclusion that the philosophy of Hegel did not apply to mathematics. There was a shift in Russell's thinking: he listened: "to a metaphysical argument for or against God or immortality or free will; and it was only after acquiring a new logic that I ceased to think such arguments worth examining."[8] Russell contrasts his father's generation with his own, the 1860s with the 1890s:

> The men who commanded respect in England in the 'sixties—Darwin, Huxley, Newman, the authors of Essays and Reviews, etc.—were not fundamentally at war with society; they could meet, as they did in the 'Metaphysical Society', to discuss urbanely whether there is a God.[9]

Ending the essay, Russell with his characteristic balance sees the pros and cons of his father's Cambridge and his own, the more violent 1890s: "The assertion of freedom seemed sufficiently noble to justify violence; the violence duly ensued, but freedom was lost in the process."[10] Freedom asserted but lost.

Russell also discusses his life in *My Mental Development*.[11] He tells how his mother died when he was two and his father when Russell was three and how he was brought up in the household of his grandparents, his grandfather being the reformer Lord John Russell. A friend of Russell's parents, John Stuart Mill, was Russell's godfather. In this chapter and in my book, I am exploring self-representation and representation. Russell is characterizing himself. Among the topics he writes about is religion:

> My father wished my brother and me to be brought up as free-thinkers, and appointed two free-thinkers as our guardians. The Court of

> Chancery, however, at the request of my grandparents, set aside the will, and I enjoyed the benefits of a Christian upbringing.[12]

Free-thinking and Christianity are recurrent themes in some of Russell's work. Besides mentioning how kind his grandfather was, as part of a portrait, Russell portrays his grandmother, including these words:

> At the age of seventy, my grandmother became a Unitarian; at the same time, she supported Home Rule for Ireland, and made friends with Irish Members of Parliament, who were being publicly accused of complicity in murder. This shocked people more than now seems imaginable. She was passionately opposed to imperialism, and taught me to think ill of the Afghan and Zulu wars, which occurred when I was about seven.[13]

This grandmother changed from being a Presbyterian to a Unitarian rather late, supported Ireland, opposed imperialism and its wars. Russell's family was involved in reform and liberalism despite his reservations about some other aspects of family life. His grandparents' household helped to form Russell: "My grandfather's library, which became my schoolroom, stimulated me in a different way."[14] Russell outlines his views in his teens:

> By the time I was fifteen, I had arrived at a theory very similar to that of the Cartesians. The movements of living bodies, I felt convinced, were wholly regulated by the laws of dynamics; therefore free will must be an illusion. But, since I accepted consciousness as an indubitable datum, I could not accept materialism, though I had a certain hankering after it on account of its intellectual simplicity and its rejection of 'nonsense'. I still believed in God, because the First-Cause argument seemed irrefutable.[15]

In his youth, Russell is considering his philosophical views, which continued to change. His description of his religious belief then disbelief is part of his self-portrait:

> I discarded first free will, then immortality; I believed in God until I was just eighteen, when I found in Mill's *Autobiography* the sentence: 'My father taught me that the question "Who made me?" cannot be answered, since it immediately suggests the further question "Who made God?" 'In that moment I decided that the First-Cause argument is fallacious.[16]

Russell thought about religion in a close and sustained way over years, so that even if he decided to be agnostic, he respected Christianity. Russell saw that there was an important part of philosophy that was connected with theology. He also talks about reading poetry, praising Shelley but not Tennyson and Byron, admiring the prose of Carlyle in *Past and Present*, but not *Sartor Resartus*; Russell says that he agreed most with Mill, whose "*Political Economy*, *Liberty*, and *Subjection of Women* influenced me profoundly."[17] Social, economic and political thought interested Russell from his youth.

Cambridge meant a great deal to Russell:

> Cambridge opened to me a new world of infinite delight. For the first time I found that, when I uttered my thoughts, they seemed to be accepted as worth considering. Whitehead, who had examined me for entrance scholarships, had mentioned me to various people a year or two senior to me, with the result that within a week I met a number who became my life-long friends.[18]

Alfred North Whitehead made a big difference in the life of Russell, who got to know people who could be friends and could share interests. These contemporaries Russell portrays as capable and hard-working but interested in matters outside academics: "poetry, philosophy, politics, ethics, indeed the whole world of mental adventure."[19] The range of culture is impressive and "mental adventure" is a felicitous phrase. Russell and his fellow students explored ideas and art outside their academic interests. He discusses these contemporaries and his teachers, his admiration for Bradley's philosophy, his "flash" recognition of the validity of the ontological argument.[20]

After Cambridge, according to Russell, he had some interesting experiences in Europe and North America. For instance, he says: "My wife was a Philadelphia Quaker, and in 1896 we spent three months in America. The first place we visited was Walt Whitman's house in Camden, N.J.; she had known him well, and I greatly admired him."[21] Russell's range of acquaintances is wide, and his admiration of Whitman is intriguing. Even if Russell is not interested in aesthetics as a field of philosophy, he reads poetry and has a sense of literary style. Russell is versatile and reflects on his options:

> In the years from 1894 to 1898, I believed in the possibility of proving by metaphysics various things about the universe that religious feeling made me think important. I decided that, if I had sufficient ability, I would devote my life to philosophy. My fellowship dissertation, on the foundations of geometry, was praised by Ward and Whitehead; if it had not been, I should have taken up economics, at which I had been working in Berlin.[22]

Religious feeling was part of his focus on metaphysics. Mathematics and philosophy were Russell's main interests, but, as he explains, he had an interest in economics. Russell gives us a map of his thinking and his development.

Russell provides an account of his move away from Hegel: "During 1898, various things caused me to abandon both Kant and Hegel. I read Hegel's *Greater Logic,* and thought, as I still do, that all he says about mathematics is muddle-headed nonsense."[23] Nor does Russell mince his words, and he also credits G. E. Moore as leading the way away from Hegel, from Bradley's view that common sense is appearance to a sense that common sense is real.[24] The relation between mathematics and philosophy was central for Russell: "Mathematics could be *quite* true, and not merely a stage in dialectic. Something of this point of view appeared in my *Philosophy of Leibniz.*"[25] At this point, Russell says that mathematics can have complete truth.

For Russell, 1900 was the most significant year in his intellectual life, which included Russell's attendance at the International Congress of Philosophy in Paris. He says: "Ever since I had begun Euclid at the age of eleven, I had been troubled about the foundations of mathematics; when, later, I came to read philosophy, I found Kant and the empiricists equally unsatisfactory."[26] Truth, certainty and foundations in mathematics and philosophy remain important for Russell. In Paris, Russell met Peano and his pupils who were mathematically precise rather than philosophically vague. Russell also says that he based himself on Peano to invent "a notation for relations" and notes that Whitehead agreed that the method was important, and soon Russell and Whitehead worked on "the definitions of series, cardinals, and ordinals, and the reduction of arithmetic to logic."[27] They had successes for a year but did not know that Frege had done much of the work. Russell maintains that his contribution to *Principia Mathematica* began "as a parenthesis in the refutation of Kant."[28] Mathematics, logic and philosophy are all connected for Russell.

In 1901, Russell says, the delight came to an end. Cantor proved there was no greatest cardinal. Russell wrote Frege who said *"die Arithmetik ist ins Schwanken geraten,"* which might be conveyed as arithmetic has begun to waver.[29] All this caused Russell despair, and he worked to no avail until 1905 when there was a glimmer. Russell's mental development is something he partly describes in terms of mathematical, logical and philosophical problems. For example, he says:

> Scholastic realism was a metaphysical theory, but every metaphysical theory has a technical counterpart. I had been a realist in the scholastic or Platonic sense; I had thought that cardinal integers, for instance, have a timeless being. When integers were reduced to classes of classes, this being was transferred to classes.[30]

Time and timelessness, metaphysics and technical philosophy preoccupied Russell. Problems he sought to solve meant that some of the solutions had errors. He thought that Meinong's work was interesting, for instance, his application of "the arguments of realism to descriptive phrases" and he considered Meinong's analysis of the sentence "'the golden mountain does not exist.'"[31] Russell focuses on language – a sentence – a matter of logic for long periods. He is a very close reader and has great analytical powers as an interpreter of words and numbers. Russell did not find Meinong's realm satisfying, so that led him to his theory of descriptions, which, he says, discovered "that, in analysing a significant sentence, one must not assume that each separate word or phrase has significance on its own account."[32] This view is interesting, and I am not sure one could say that in analysing a poem or poetry, for instance, the word "be" or the phrase "To be or not to be" in Hamlet's soliloquy. The difference between "the" and "an" can be significant – certainly in a poem but perhaps in any phrase or sentence. Russell may well mean something different. What does Russell mean? He elaborates:

> 'The golden mountain' can be part of a significant sentence, but is not significant in isolation. It soon appeared that class symbols could be treated like descriptions, i.e. as non-significant parts of significant sentences. This made it possible to see, in a general way, how a solution of the contradictions might be possible. The particular solution offered in *Principia Mathematica* had various defects, but at any rate it showed that the logician is not presented with a complete impasse.[33]

It seems that Russell is talking from the point of view of a logician, whereas others may read sentences in terms of literary, metaphysical and other forms of analysis. He recognizes the limitations of his thinking or work in his greatest work in the field of mathematical logic, but he also tries to see what his contribution might be, and that is avoiding a full impasse.

Russell is good at expressing his mental development in terms of problems he faced and solutions or theories he proposed. He explains further about his theory:

> The theory of descriptions, and the attempt to solve the contradictions, had led me to pay attention to the problem of meaning and significance. The definition of 'meaning' as applied to words and 'significance' as applied to sentences is a complex problem, which I tried to deal with in *The Analysis of Mind* (1921) and *An Inquiry into Meaning and Truth* (1940).[34]

His work on descriptions led him to that on meaning and significance. One problem leads to another problem often through a possible but limited solution that never seems to satisfy. Contradictions cry out for resolution but may lead to more contradictions. The problem of meaning and significance, according to Russell, leads to psychology and physiology. Russell says that he came to question "the complete independence of logic."[35] He considers logic to be a more exact science than psychology, so that he thinks it desirable "to delimit the problems that can be dealt with by logical methods."[36] In this context, Russell uses Occam's razor, which was metaphysical, "a principle of parsimony as regards 'entities,'" and which was the way Russell thought about it when he was writing *Principia Mathematica.*[37] Russell discusses a "minimum vocabulary," which he defines as "one in which no word can be defined in terms of the others."[38] He has many interesting things to say on this subject. For instance, Russell observes: "There is need, however, of great caution in drawing inferences from minimum vocabularies. In the first place, there are usually, if not always, a number of different minimum vocabularies for a given subject-matter."[39] Russell often offers qualifications and limitations, limiting the claims of his own work and that of others. He discusses one type of inference from studying minimum vocabularies, for example, the problem of universals, no vocabulary being able to dispense with words of the kind deemed "universals."[40] While stating theories and problems, Russell is also good at providing examples. Here in the context of universals, he focuses on one word: "But the fact that we need the word 'similar' indicates some fact about the world, and not only about language. What fact it indicates about the world, I do not know."[41] The relation of word and world occupies Russell, and he also admits the limits of his own knowledge. Furthermore, Russell discusses minimum vocabularies in relation to historical events. He also talks about the theory of knowledge, a key concern for him, as being subjective and asking "'how do I know what I know?'" and thus beginning "from personal experience."[42] The problems of philosophy are keys to the growth of Russell's mind. He sets these matters out clearly:

> I am not a solipsist, nor an idealist; I believe (though without good grounds) in the world of physics as well as in the world of psychology. But it seems clear that whatever is not experienced must, if known, be known by inference. I find that the fear of solipsism has prevented philosophers from facing this problem, and that either the necessary principles of inference have been left vague, or else the distinction between what is known by experience and what is known by inference has been denied.[43]

Denying solipsism and idealism, Russell speaks about his belief in physics and psychology and how what is not known through experience must be known through inference but adds that philosophers have feared solipsism and that they have denied distinction between knowledge by experience and that by inference. Belief is something that is unavoidable even if it cannot be proven. Russell also says what would be worth working on:

> If I ever have the leisure to undertake another serious investigation of a philosophical problem, I shall attempt to analyse the inferences from experience to the world of physics, assuming them capable of validity, and seeking to discover what principles of inference, if true, would make them valid. Whether these principles, when discovered, are accepted as true, is a matter of temperament; what should not be a matter of temperament should be the proof that acceptance of them is necessary if solipsism is to be rejected.[44]

If he had the time, he would examine inference from experience to physics with the assumption of validity and the connection among principle, truth and proof. Self-regard is not a good option.

Although Russell focused most on mathematics and philosophy, he thought and wrote about other subjects, and he explains why in this autobiographical essay. He frames these topics as part of "social questions."[45] Russell grew up in a political household but says he preferred philosophy. None the less, he studied and wrote about politics: "In 1895, when in Berlin, I made a study of German Social Democracy, which I liked as being opposed to the Kaiser, and disliked as (at that time) embodying Marxist orthodoxy."[46] Russell gives us a balanced view. He says Sidney Webb influenced him to be an imperialist, and Russell supported the Boar War, but in 1901 he abandoned this view and henceforth "felt an intense dislike of the use of force in human relations, though I always admitted that it is sometimes necessary."[47] Russell shows how he has grown and changed his mind and does not stick to a position he comes to see as mistaken. He says that, as an internationalist, he spoke up against Joseph Chamberlain for turning against free trade in 1903. Russell actively agitated for women's suffrage. In 1910, when Russell wished to stand for Parliament, the selection committee discovered he was a free thinker. During the First World War, Russell wrote about preventing future wars, and this made him known more widely, but the Treaty of Versailles destroyed this hope for future peace.[48] Russell contrasts the Soviet Union with China:

> Many of my friends saw hope in Soviet Russia, but when I went there in 1920 I found nothing that I could like or admire. I was then invited to China, where I spent nearly a year. I loved the Chinese, but it was obvious that the resistance to hostile militarisms must destroy much of what was best in their civilization. They seemed to have no alternative

> except to be conquered or to adopt many of the vices of their enemies. But China did one thing for me that the East is apt to do for Europeans who study it with sensitive sympathy: it taught me to think in long stretches of time, and not to be reduced to despair by the badness of the present. Throughout the increasing gloom of the past twenty years, this habit has helped to make the world less unendurable than it would otherwise have been.[49]

Whereas Russell could not admire Soviet Russia when he went there, he "loved the Chinese" when he lived there for almost a year and he saw the external militarism and vices, but he learned from China how to take the long view and endure rather than despair in the present. Russell sought hope. He explored other cultures, and here he found Chinese culture taught him something important.

Education also concerned Russell. He explains that after China, his two elder children were born, which interested him in early education:

> I have been supposed to be an advocate of complete liberty in schools, but this, like the view that I am an anarchist, is a mistake. I think a certain amount of force is indispensable, in education as in government; but I also think that methods can be found which will greatly diminish the necessary amount of force.[50]

Through self-characterization, Russell tries to correct how some characterize him. He does not favour force but sees that some force is sometimes necessary. Russell elaborates:

> The question of discipline in childhood, like all other practical questions, is one of degree. Profound unhappiness and instinctive frustration is apt to produce a deep grudge against the world, issuing, sometimes by a very roundabout road, in cruelty and violence.[51]

Russell is interested in social and psychological problems and considered them in the First World War. That more than a few enjoyed the war, he views as a failure of education. He sees the role of parents and the state in education, the state being involved in economic and political reforms. Russell shifts back to the personal:

> The world, however, was moving more and more in the direction of war and dictatorship, and I saw nothing useful that I could do in practical matters. I therefore increasingly reverted to philosophy, and to history in relation to ideas.[52]

For Russell, when he could not make a practical contribution to a darkening world, he turned to philosophy and the history of ideas.

Three subjects occupied Russell most:

> History has always interested me more than anything else except philosophy and mathematics. I have never been able to accept any general schema of historical development, such as that of Hegel or that of Marx. Nevertheless, general trends can be studied, and the study is profitable in relation to the present.[53]

Russell finds uses for history as it can help study past trends for the present. He also thinks about philosophy in context: "The relation of philosophy to social conditions has usually been ignored. by professional philosophers. Marxists are interested in philosophy as an effect, but do not recognize it as a *cause*. Yet plainly every important philosophy is both."[54] Plato is an example Russell provides. He sees Plato partly "an effect of the victory of Sparta in the Peloponnesian war, and is also in part among the causes of Christian theology" and observes that to consider Plato only as that effect "is to make the growth of the medieval church inexplicable."[55] Russell says that he is writing "a history of western philosophy from Thales to the present" and that he treats each significant system "as an effect and as a cause of social conditions."[56] Thus, Russell writes the history of philosophy with a regard to social causes and effects.

The frankness, balance and self-criticism or self-limitation are part of the appeal of Russell's autobiographical writing. Russell presents a history of his own ideas. He represents his feelings, thoughts, aspirations, disappointments and much else. His style is clear and appealing:

> My intellectual journeys have been, in some respects, disappointing. When I was young I hoped to find religious satisfaction in philosophy; even after I had abandoned Hegel, the eternal Platonic world gave me something nonhuman to admire. I thought of mathematics with reverence, and suffered when Wittgenstein led me to regard it as nothing but tautologies.[57]

Russell speaks of some aspects of the disappointment of his "intellectual journeys," his youthful hope for "religious satisfaction in philosophy," his admiration of "the eternal Platonic world" even after his abandonment of Hegel, his reverence for mathematics until Wittgenstein caused Russell to see "it as nothing but tautologies."[58] Loss of faith and admitting the dropping of earlier ideas or beliefs is a recurring theme in Russell's autobiographic portrait. Hope leads to disappointment, but Russell does not seem to give up all hope, but he limits it. He changes his mind. He looks for evidence. He seeks truth and then sees the limits of claims to truth.

Being a skilled writer and a philosopher who chooses his words carefully, Russell brings together form and content in a distinctive way, so much so that I feel obliged to quote him often and at some length when presenting my analysis:

> I have always ardently desired to find some justification for the emotions inspired by certain things that seemed to stand outside human life and to deserve feelings of awe. I am thinking in part of very obvious things, such as the starry heavens and a stormy sea on a rocky coast; in part of the vastness of the scientific universe, both in space and time, as compared to the life of mankind; in part of the edifice of impersonal truth, especially truth which, like that of mathematics, does not merely describe the world that happens to exist.[59]

One might expect an analytical philosopher – a logician and mathematician – to talk about reason, but Russell also uses words and phrases of feeling as in the following: "ardently desired," "emotions inspired" and "feelings of awe." He invokes scenes or images of heavens, sea and universe much greater than humans with "an impersonal truth" like that of mathematics. Here the person meets the impersonal. He then discusses those who would make humanism into a religion, recognizing nothing that is greater than humans, and he notes that such a view does "not satisfy my emotions."[60] Russell shows the tension between his intellect and feelings, and his analysis is frank, artful, sincere and arresting. He presents his qualification to his qualification of humanism:

> And yet I am unable to believe that, in the world as known, there is anything that I can value outside human beings, and, to a much lesser extent, animals. Not the starry heavens, but their effects on human percipients, have excellence; to admire the universe for its size is slavish and absurd; impersonal non-human truth appears to be a delusion. And so my intellect goes with the humanists, though my emotions violently rebel. In this respect, the 'consolations of philosophy' are not for me.[61]

Russell uses the word "believe" here and elsewhere. With some reluctance, he comes to the conclusion that he can only value humans rather than the heavens, the universe and impersonal truth. He says that his intellect is with the humanists even as his emotions rebel violently, so that he says, in this regard, for him there are no "'consolations of philosophy." This seems like an allusion to Boetius' book of that title in which the female allegorical figure of Philosophy consoles the autobiographical figure of Boethius' book awaiting execution (Boethius wrote the book waiting to be executed) and holds out happiness, arguing for virtues and for good and, borrowing from Plato, sees that happiness is divine or God.[62]

After discussing his disappointments, Russell has found satisfaction in philosophy in "intellectual ways."[63] He is glad that matters vague in his youth are open "to an exact technique," which allows for the type of progress made in science.[64] Russell says that sometimes it is possible to prove that definite knowledge cannot be attained and that where definite knowledge is unattainable, it is sometimes possible to prove that hypotheses based on evidence can be formulated. For Russell, philosophers using methods "from logical analysis" are able to have arguments among themselves in a cooperative way and can argue with each other and agree on "the outcome."[65] This, Russell observes, occurred in his life, and Frege, he acknowledges, pioneered the method. Russell values this extension of reason. Using the Chinese long view that he mentioned before, Russell says that war and superstition may choke philosophical reason, "but one may hope that it will not be lost utterly or for more than a few centuries."[66] In the last sentence of the essay, Russell affirms the happiness of his philosophical life in intellectual terms.

Russell published some of his memories in 1956 that were later included in *Basic Writings*.[67] Some of the memories are similar to those in the autobiographical writing I have been analysing, so I will discuss them accordingly. This piece opens with an appeal to the reader:

> For those who are too young to remember the world before 1914, it must be difficult to imagine the contrast for a man of my age between childhood memories and the world of the present day. I try, though with indifferent success, to accustom myself to a world of crumbling empires, Communism, atom bombs, Asian self-assertion, and aristocratic downfall. In this strange insecure world where no one knows whether he will be alive tomorrow, and where ancient states vanish like morning mists, it is not easy for those who, in youth, were accustomed to ancient solidities to believe that what they are now experiencing is a reality and not a transient nightmare.[68]

For Russell, 1914 or the beginning of the First World War was the turning point, perhaps a wound, which is what trauma meant in ancient Greek. One of the themes in Russell's own depiction of his life is a desire for certainty and security, which is all too human, and perhaps amplified from losing his parents so young. Like Winston Churchill, who lost his father when young but not as young as Russell was, Russell repairs the ruin and redeems the loss by remembering and writing about his parents (Russell and his wife of the time edited the works of his parents). Death came early, and Russell affirms life in so many ways. He tries to find faith in something in a hostile world in which nature has robbed him of his parents. I am certainly not reducing Russell to this observation, hypothesis or conjecture, but I am suggesting it might have been one contributing factor in his life and intellectual life. One should look at social context but not reduce a person's writing to that milieu. He sought "solidities" that vanished. This vanishing is something I have remarked on in Russell's life and mental adventure. He desires something solid, certain, provable, to believe, but despite his great efforts and skill, the world and the truth can be unsure, where Russell cannot prove the existence or non-existence of God, wishes for Plato's eternal realm but discovers that mathematics is tautological. The uncertainty of the atomic age, the possibility of destruction at any time, is so unlike the world before 1914. His family, aristocrats working for freedom and democracy since the late seventeenth century, finds itself in a strange world not apparently in need of such aristocrats. Russell being Russell would be aware of all this. He knows the trauma and the scars and returns to the First World War in his writing: it is the breach in his life that affects him most after that of the loss of his parents.

Tradition is no more. Barely remembering his parents, although this does not mean that Russell cannot retrospectively miss and mourn them, Russell turns to a portrait of his grandfather, the great reformer, Lord John Russell. He says that his grandfather was born early in the French Revolution, served in parliament during the time of Napoleon, was a Whig who followed Fox and thought the English too hostile to the revolution in France and to Napoleon, whom he saw in exile in Elba. Russell notes that his grandfather's introduction of the Reform Bill of 1832 began the road for England to democracy and that during the revolutions of 1848, he was Prime Minister. For Russell, Lord John inherited from his ancestor in the Revolution of 1688, and "aristocratic liberalism."[69] Russell explains what his family taught him in this regard: "I was taught a kind of theoretic republicanism which was prepared to tolerate a monarch so long as he recognized that he was an employee of the people and subject to dismissal if he proved unsatisfactory."[70] Colonial Rhode Island and Connecticut especially seem to have fostered a view of monarchy not dissimilar to what Russell describes. The king was the chief magistrate. I am quite sympathetic to Russell's family's attempt to reform for democracy and freedom. Although we carry with us a myth or a feeling of the golden age, especially if we had happy childhoods or found university to be a liberation – as I also did – we may come to the conclusion that there is no golden age and that history, to paraphrase Joyce's Stephen Dedalus, is a nightmare from which we are trying to awaken. A golden age may well be an oxymoron, yet another tautology. Apparently, as Russell says, his grandfather used to explain this view to Queen Victoria, who, despite her lack of sympathy for it, provided him a house in Richmond Park where Russell spent his youth. Russell says that he has retained political principles but not expectations and sums up the matter: "There was to be ordered progress throughout the world, no revolutions, a gradual cessation of war, and an extension of parliamentary government to all those unfortunate regions which did not yet enjoy it."[71] Russell's wit and humour poke some fun at the worldview of his grandparents' liberal household. For instance, he says:

> My grandmother used to laugh about a conversation she had had with the Russian Ambassador: she said to him, 'Perhaps some day you will have a parliament in Russia', and he replied, 'God forbid, my dear Lady John.' The Russian Ambassador of today might give the same answer if he changed the first word. The hopes of that period seem now a little absurd. There was to be democracy, but it was assumed that the people would always be ready to follow the advice of wise and experienced aristocrats.[72]

While Russell is nostalgic for the time before 1914, he does look back at its absurdities. He remembers his grandparents and their ideas fondly but also understands the changes in the world. Politics and religion were central aspects of the household. Politics had its twists and turns:

> It might perhaps be admitted also that Cromwell had gone too far in cutting off the king's head but, broadly speaking, anything done against kings was to be applauded—unless, indeed, it were done by priests, like Becket, in which case one sided with the king.[73]

With wit, Russell likes to show the contradictory and the incongruous or at the very least the subtleties of positions, ideas, beliefs. Here is a religious household that will not side with priests. Pious observance was part of Russell's youth: "The atmosphere in the house was one of puritan piety and austerity. There were family prayers at eight o'clock every morning."[74] Russell experienced pressure to follow in the family tradition of politics and grew up in a liberal but religious aristocratic household. The priority there Russell describes: "Only virtue was prized, virtue at the expense of intellect, health, happiness, and every mundane good."[75] Virtue above all. Russell outlines his intellectual rebellion against such an atmosphere. His family suspected mathematics because it lacked ethical content and did not prize philosophy, for which they had a joke: "'What is mind? No matter. What is matter? Never mind.'"[76] This was witty but wore on Russell after much repetition.

As elsewhere, Russell stresses the importance of Cambridge to him: "When at the age of eighteen I went up to Cambridge, I found myself suddenly and almost bewilderingly among people who spoke the sort of language that was natural to me."[77] He portrays his move from a morbid home with an unwholesome morality to a place where intelligence was prized. Cambridge freed him: "And to find myself in a world where intelligence was valued and clear thinking was thought to be a good thing caused me an intoxicating delight."[78] He says that he was no longer lonely. Russell describes his interests in philosophy and mathematics at Cambridge and says of the latter:

> I was especially attracted to problems concerning the foundations of mathematics. I wished to believe that some knowledge is certain and I thought that the best hope of finding certain knowledge was in mathematics. At the same time it was obvious to me that the proofs of mathematical propositions which my teachers had offered me were fallacious. I hoped that better proofs were forthcoming. Subsequent study showed me that my hopes were partly justified. But it took me nearly twenty years to find all the justification that seemed possible and even that fell far short of my youthful hopes.[79]

Once more, Russell discusses the problems of mathematics and how many years he worked and how even then he did not find the justifications he had hoped for as a student. Russell's self-portrait is one of hope and disappointment, achievement and limitation.

As Russell states elsewhere, his central choice was between politics – his family tradition – and philosophy. Russell says he found the conflict with his family painful when he resisted their pressure and chose philosophy. He preferred peace but followed his beliefs, as he has done since:

> But over and over again profound convictions have forced me into disagreements, even where I least desired them. After I had decided on philosophy, however, everything went smoothly for a long time. I lived mainly in an academic atmosphere where the pursuit of philosophy was not regarded as an eccentric folly. All went well until 1914.[80]

The wound, the reversal, the catastrophe of the First World War recur in Russell's autobiographical writing. At Cambridge, he could pursue philosophy in peace until the great conflict came. Russell provides his version of the tragedy of the First World War similar to the one I have discussed. What he stresses here is how he fell out with his friends over his view that England should remain neutral in the war, and this estranged him from most of his former friends and from national life. Russell was surprised at his own strength: "But something that if I had been religious I should have called the Voice of God, compelled me to persist."[81] Russell often uses religious imagery and had a religious upbringing, so that even when he turned away from that, the impulse, even if made secular, remains. He clarifies his position on war: "Neither then nor later did I think *all* war wrong. It was that war, not all war, that I condemned."[82] The First World War had a great effect:

> We owe to the first war and its aftermath Russian Communism, Italian Fascism and German Nazism. We owe to the first war the creation of a chaotic unstable world where there is every reason to fear that the Second World War was not the last, ...[83]

He enumerates many ill effects: "All these evils have sprung with the inevitability of Greek tragedy out of the First World War."[84] Russell, who uses a literary comparison in setting out the tragic fate of the war, does not lose this view of the mistakes and aftermath of 1914. The point he makes elsewhere, he reiterates here:

> The Kaiser's Germany, although war propaganda on our side represented it as atrocious, was in fact only swashbuckling and a little absurd. I had lived in the Kaiser's Germany and I knew that progressive forces in that country were very strong and had every prospect of ultimate success. There was more freedom in the Kaiser's Germany than there is now in any country outside Britain and Scandinavia. We were told at the time that it was a war for freedom, a war for democracy and a war against militarism. As a result of that war freedom has vastly diminished and militarism has vastly increased. As for democracy, its future is still in doubt.[85]

Russell often changes his mind when he thinks he is wrong, in mathematics and in other matters, for instance, regarding imperialism, but here he holds the same position. He repeats that English neutrality in the First World War would have been in a better state as it would have allowed Germany a victory in short order. Russell says that he still holds this view and does not regret trying in the First World War to counter the official propaganda about the wickedness of the Germans, which led to the "evil" that came out of the "severity" of the Treaty of Versailles.[86]

This is alternative history, so we can never know the what if of another road taken in history. One might have to go a way to see whether the Anglo-French understanding, such as in the Entente Cordiale of 1904, was a mistake and why the series of agreements were signed in the first place. Would it be easy for Britain to break its word on treaties and remain neutral? In any case, the propaganda was distasteful, the Treaty of Versailles harsh, and neither would help the cause of peace, prosperity and democracy. The next two chapters will examine the representations of Alan Turing.

Notes

1 Bertrand Russell, "My Religious Reminiscences," in *The Basic Writings of Bertrand Russell*, edited by Robert E. Egner and Lester E. Denonn with an introduction by John G. Slater, 1st ed. (London: Routledge, 2009), 3–8. This essay appeared first in *The Rationalist Annual*, 1938, published by C. A. Watts & Co., Ltd.
2 Russell, "My Religious Reminiscences," 3; see 4–8. On Russell, see, for instance: Alan Wood, *Bertrand Russell: The Passionate Sceptic* (London: Allen and Unwin, 1957); Joe Park, *Bertrand Russell on Education* (Columbus: Ohio State University Press, 1963); BY A STAFF REPORTER, "Bertrand Russell Dies at 97 After Flu," *Times*, 3 February 1970, 1. The Times Digital Archive, link.gale.com/apps/doc/CS17920579/TTDA?u=camb55135&sid=bookmark-TTDA&pg=1&xid=4cb67873; Trevor Fishlock, "The World Pays Tribute to Bertrand Russell," *Times*, 4 February. 1970, 9. The Times Digital Archive, link.gale.com/apps/doc/CS151220804/TTDA?u=camb55135&sid=bookmark-TTDA&pg=9&xid=5996d5fa; Ronald William Clark, *The Life of Bertrand Russell* (London: Jonathan Cape and Weidenfeld & Nicolson, 1975); Katharine Tait, *My Father Bertrand Russell* (New York: Harcourt Brace Jovanovich, 1975); I. F. Stone, "Bertrand Russell as a Moral Force in World Politics," *Russell* 1 (1981): 7–25; Caroline Moorehead, *Bertrand Russell* (New York: Viking, 1992); Bart Schultz, "Bertrand Russell in Ethics and Politics," *Ethics* 102 (1992): 594–634; Stefan Andersson, *In Quest of Certainty* (Stockholm: Almqvist & Wiksell, 1994); Michael K. Potter, *Bertrand Russell's Ethics* (London and New York: Continuum Books, 2006); Maya Kronfeld, "'Prufrock' between Acquaintance and Description: Bertrand Russell and T. S. Eliot," *Philosophy and Literature* 47 (1) (2023), 167–183; Normand Baillargeon et Chantel Santerre, *Bertrand Russell* (Paris: Presses Universitaires de France, 2025). On the vote versus the bag of grain and the relation between desire and duty and his witty comment on the desire to duteous, see Bertrand Russell – Nobel Prize Acceptance Speech (1950), https://www.youtube.com/watch?v=HzTl-eS8NGw#ddg-play.
3 Russell, "My Religious Reminiscences," 4.
4 Russell, "My Religious Reminiscences," 4.
5 Russell, "My Religious Reminiscences," 5.
6 Russell, "My Religious Reminiscences," 6–7.
7 Russell, "My Religious Reminiscences," 7.
8 Russell, "My Religious Reminiscences," 7.
9 Russell, "My Religious Reminiscences," 7–8.
10 Russell, "My Religious Reminiscences," 8.
11 Bertrand Russell, "My Mental Development," in *Basic Writings of Bertrand Russell*, edited by Robert E. Egner and Lester E. Denonn with an introduction by John G. Slater, 1st ed. (London: Routledge, 2009), 9–22. The original appeared in *The Philosophy of Bertrand Russell*, edited by Paul Arthur Schilpp, Library of Living Philosophers (New York: Tudor Publishing Co., 1951)
12 Russell, "My Mental Development," 10.
13 Russell, "My Mental Development," 11.
14 Russell, "My Mental Development," 11.
15 Russell, "My Mental Development," 12.
16 Russell, "My Mental Development," 12–13.
17 Russell, "My Mental Development," 13.
18 Russell, "My Mental Development," 13.

19 Russell, "My Mental Development," 13.
20 Russell, "My Mental Development," 14.
21 Russell, "My Mental Development," 15.
22 Russell, "My Mental Development," 15.
23 Russell, "My Mental Development," 15.
24 Russell, "My Mental Development," 15.
25 Russell, "My Mental Development," 16.
26 Russell, "My Mental Development," 16.
27 Russell, "My Mental Development," 16.
28 Russell, "My Mental Development," 16.
29 Russell, "My Mental Development," 16.
30 Russell, "My Mental Development," 16–17.
31 Russell, "My Mental Development," 17.
32 Russell, "My Mental Development," 17.
33 Russell, "My Mental Development," 17.
34 Russell, "My Mental Development," 17.
35 Russell, "My Mental Development," 17.
36 Russell, "My Mental Development," 17.
37 Russell, "My Mental Development," 17.
38 Russell, "My Mental Development," 17.
39 Russell, "My Mental Development," 18.
40 Russell, "My Mental Development," 18.
41 Russell, "My Mental Development," 18.
42 Russell, "My Mental Development," 19.
43 Russell, "My Mental Development," 19.
44 Russell, "My Mental Development," 19.
45 Russell, "My Mental Development," 19.
46 Russell, "My Mental Development," 19.
47 Russell, "My Mental Development," 19.
48 Russell, "My Mental Development," 19–20.
49 Russell, "My Mental Development," 20.
50 Russell, "My Mental Development," 20.
51 Russell, "My Mental Development," 20.
52 Russell, "My Mental Development," 20–21.
53 Russell, "My Mental Development," 21.
54 Russell, "My Mental Development," 21.
55 Russell, "My Mental Development," 21.
56 Russell, "My Mental Development," 21.
57 Russell, "My Mental Development," 21.
58 Russell, "My Mental Development," 21.
59 Russell, "My Mental Development," 21.
60 Russell, "My Mental Development," 21.
61 Russell, "My Mental Development," 21.
62 Boethius, *Theological Tractates. The Consolation of Philosophy*, translated by H. F. Stewart, E. K. Rand, S. J. Tester, Loeb Classical Library 74 (Cambridge, MA: Harvard University Press, 1973); see John Marenbon, "Anicius Manlius Severinus Boethius", *The Stanford Encyclopedia of Philosophy* (Winter 2021 Edition), Edward N. Zalta (ed.), https://plato.stanford.edu/archives/win2021/entries/boethius/.
63 Russell, "My Mental Development," 22.
64 Russell, "My Mental Development," 22.

65 Russell, "My Mental Development," 22.
66 Russell, "My Mental Development," 22.
67 Bertrand Russell, "Adaptation: An Autobiographical Epitome," in *Basic Writings of Bertrand Russell*, edited by Robert E. Egner and Lester E. Denonn with an introduction by John G. Slater, 1st ed. (London: Routledge, 2009), 23–27. Published originally in *Portraits from Memory* (London: Allen & Unwin; New York: Simon & Schuster, 1956).
68 Russell, "Adaptation," 23.
69 Russell, "Adaptation," 23.
70 Russell, "Adaptation," 23.
71 Russell, "Adaptation," 24.
72 Russell, "Adaptation," 24.
73 Russell, "Adaptation," 24.
74 Russell, "Adaptation," 24.
75 Russell, "Adaptation," 25.
76 Russell, "Adaptation," 25.
77 Russell, "Adaptation," 25.
78 Russell, "Adaptation," 25.
79 Russell, "Adaptation," 25.
80 Russell, "Adaptation," 26.
81 Russell, "Adaptation," 26.
82 Russell, "Adaptation," 26.
83 Russell, "Adaptation," 26.
84 Russell, "Adaptation," 26.
85 Russell, "Adaptation," 26–27.
86 Russell, "Adaptation," 27.

10 Representing Alan Turing and the Making of a Myth

Alan Turing was a person and not a myth, but eminent people soon become myths even if they are humble, other-worldly or just interested in making something or solving a problem far from the madding crowd. My focus in this book has been, especially in relation to scientists, self-representation and representation. Whereas Darwin wrote something about himself, Pierre Curie and Alan Turing did not set out to write a memoir or autobiography, and nothing like that survives if ever such a text were written. Instead, in the wake of the deaths of Pierre Curie and Alan Turing, we have what the family says, Marie Curie's account of Pierre, and Turing's mother's biography, both brief but both written in the wake of sudden deaths at relatively young ages, of Pierre and Alan, in their forties, here one day and gone the next. Both Marie Curie and Sarah Turing came to write of their respective traumas and did so in some way to remember and celebrate the character and genius of the men they had been so close to in their families. Marie was a scientist and worked with Pierre, collaborated with him and understood the scientific problems. The Curies shared mathematics, physics and chemistry. Sarah was not a mathematician or logician. Both women expressed their love and support by telling the stories of those men who left them prematurely. Both had gaps in their narratives and had to rely on others for times they were not present in the lives of their subjects. In Sarah Turing's case, and this is my focus here, I shall follow what I have done throughout the book, and that is, for the most part, a chronological analysis. Thus, I begin with Sarah Turing's biography of Alan Turing and then proceed to a briefer discussion of Alan and an elucidation and perhaps even an attempt at correction of that biography by John Turing (1908–1983), a solicitor, written in the 1960s and 1970s (and perhaps the early 1980s) and published in 2012 in the volume published by Cambridge University Press I am focusing on here. As I am concentrating on families in the self-representation and representation of scientists in this part of my book, I then discuss briefly aspects of comments by Andrew Hodges, a mathematician at Oxford and a biographer of Turing who understood

DOI: 10.4324/9781003651871-11

his mathematics, especially how it relates to Sarah Turing's representation of her son in her brief but influential biography, the one I am focusing on most in my discussion in this chapter.

But first let me tell you how I came back to Alan Turing. My good friend and mathematician, Anders Lindquist, a colleague at Shanghai Jiao Tong University, and I were planning to write an article on artificial intelligence and part of what we discussed was Alan Turing and the problem of the Chinese room. Anders sent me a few paragraphs to consider; he being accomplished at the highest levels of mathematics, physics and engineering and I, although very interested in mathematics and physics in school, had turned to poetry, history of ideas and science and culture in literature and history. Anders knows more about humanities than I do about mathematics, but we thought about artificial intelligence in terms of ethics and related matters. That is our common interest. Here are the paragraphs he sent me by email on 19 November 2024:

> Our world has gone through four industrial revolutions. The first came with mechanization, waterpower, and in 1782 with the steam engine. The second introduced mass production, electricity, and assembly lines with conveyor belts in 1913. The third brought automation and the first computers in the 1950s. We are now in the fourth industrial revolution with cyber physical systems enabling intelligent automation.
>
> Cyber physical systems are engineered systems which seamlessly connect computational algorithms and a physical component. The cyber part performs mathematical procedures, the results of which are transferred to the physical part via actuators, and then sensors bring back the state of the physical system to the cyber physical system, thus creating a feedback loop. The cornerstones of intelligent manufacturing are advanced robotics, big-data processing, communication between manufacturing devices, networks achieved through internet connectivity, communication between manufacturing devices, and in particular artificial intelligence.
>
> The concept of artificial intelligence, i.e., intelligence exhibited by machines, rather than humans, was introduced by Alan Turing (1912–1954), who formalized the concepts of algorithm. He introduced the abstract concept of the Turing machine, which became a model of a general-purpose computer. Turing posed the question whether machines can have intelligence, which became an important philosophic question, to be further discussed below. This is the strong AI hypothesis that an appropriately programmed computer could have a mind in the same sense human beings have minds. This is related to the classical concept of Mechanism, a classical philosophical question whether living things are similar to complicated machines. Related concepts are Cybernetics,

> the scientific study of control and communication in animal and machine (Norbert Wiener, 1948) and Machine learning, i.e. giving computers the ability to learn without being explicitly programmed. (Arthur Samuel, 1959)[1]

This lucid account was part of his contribution to our exchange on the Chinese room, which was related to Turing and Professor Lindquist's and my interest in artificial intelligence or machine learning.

In 1980, John Searle published "Minds, Brains and Programs" in the journal *The Behavioral and Brain Sciences* and in 1984 set out the Chinese Room argument in *Minds, Brains and Science.* Searle imagines himself following a symbol-processing programme in English, what Turing called "a paper machine." In this imagined scene, Searle follows English instructions for manipulating Chinese symbols, whereas a computer is subject to a programme written in a computing language. Searle and the computer appear to understand Chinese. This is a thought experiment that constitutes part of Searle's argument and is known as the Chinese Room Argument. Neither Searle nor the computer understands Chinese but each follows the programme; he sends Chinese characters under the door, and others outside think there is a Chinese speaker in the room. Searle concludes that programming a computer only makes it appear to understand, so that the "Turing Test" is not adequate and human and computer minds are not the same. In 1999, Searle summarizes the Chinese Room Argument:

> Imagine a native English speaker who knows no Chinese locked in a room full of boxes of Chinese symbols (a data base) together with a book of instructions for manipulating the symbols (the program). Imagine that people outside the room send in other Chinese symbols which, unknown to the person in the room, are questions in Chinese (the input). And imagine that by following the instructions in the program the man in the room is able to pass out Chinese symbols which are correct answers to the questions (the output). The program enables the person in the room to pass the Turing Test for understanding Chinese but he does not understand a word of Chinese.[2]

Searle's summary, especially the last sentence, goes to the nub of the matter, and in 2010, he helps to elucidate further:

> I demonstrated years ago with the so-called Chinese Room Argument that the implementation of the computer program is not by itself sufficient for consciousness or intentionality (Searle 1980). Computation is defined purely formally or syntactically, whereas minds have actual mental or semantic contents, and we cannot get from syntactical to the

> semantic just by having the syntactical operations and nothing else. To put this point slightly more technically, the notion "same implemented program" defines an equivalence class that is specified independently of any specific physical realization. But such a specification necessarily leaves out the biologically specific powers of the brain to cause cognitive processes. A system, me, for example, would not acquire an understanding of Chinese just by going through the steps of a computer program that simulated the behavior of a Chinese speaker.
>
> (p. 17)[3]

Once more, Searle stresses that simulation is not the same as consciousness. Does a computer have motive or intention? The passage contrasts human understanding with computer simulation.

The Chinese Room Argument is the idea of a paper machine, a computer implemented by a human. In "Intelligent Machinery" (1948), Turing wrote a programme for a "paper machine" to play chess, a paper machine being a type of programme, a series of simple steps written in language – English – that a human implements. The human operator of this machine follows instructions and need not know chess. Turing thought computers could be part of conversations and behave intelligently and proposed a test (the Turing Test) – if in a conversation a computer could pass for human, it should be regarded as intelligent.[4] My own simple question is "Can computers feel?" How would the human feel if he or she found out that it was a computer on the other side? Embarrassment, shame or foolishness for being duped? How would the computer feel if it discovered that the human could not really speak Chinese? It could not feel. It might simulate a feeling it was programmed to express as if it were a thought about a feeling. Human poetry, for instance, is a representation of feeling and not just thought from a human poet to a human reader based on human experience. The poet expresses through a poem a human situation (thought, feeling, words, experience) to another human. The poem might be simulated, but a poem written by a computer and read by a human or a poem composed by a computer for another computer is a simulation of thought and feeling, so while Turing helped to prompt this debate, which is a valuable debate, I tend more to Searle's view. But that got me to go back to Turing because I knew Searle's work better, and Searle was responding to Turing, and Professor Lindquist and I were thinking about the origins of artificial intelligence and the role of algorithms.

That was a digression about my motivation and interest in Turing in recent years. Here I shall respond to the fine Centenary edition of 2012, not as it is presented, but in chronological order, so that I shall present my response to Sara Turing's representation of her son in the biography of 1959 and then proceed to the foreword to that biography, published by Heffers, to John Turing's response to the biography from the 1960s. I shall then examine briefly Andrew Hodge's "Author's Note" in 1983, his response to the biography before returning to the 2012 edition and then to Martin Davis' foreword to that book. This is part of my training and how I work, so rather than duplicate the work of others, I respond to the original work and see how its reception unfolds. I try to avoid filters but also respect the editorial process. The original book of 1959 had a small print run and is not readily available, or I would have started with that as a book.

This first edition of 1959 is at the heart of the second edition of 2012. E.S.T. or Sara Turing writes a Preface to the First Edition, which sets out the goal and structure of her book:

> The aim of this book is to trace from early days the development of a mathematician and scientist of great originality and to record details from which a selection may be made by a future biographer. Owing to the enforced silence regarding my son's activities in the Foreign Office during the Second World War, there is, except for some few anecdotes, a regrettable gap of six years in the narrative. The book is divided into two parts. The former and major part is mainly biographical but contains sufficient scientific material to indicate the scope and depth of my son's research. The second part affords more technical particulars which might prove wearisome to the general reader: but these barely touch the fringe of his work on Computing Machines and Morphogenesis.[5]

It is the first part that is of most interest to me here because the second part involves repetition and sets out responses from experts on technical matters, but that is less important to my analysis of Sara Turing's representation of Alan Turing. The Foreword by Lyn Irvine, married to Max Newman, was the second text after the one-page Preface of Sara Turing. Irvine's Foreword to the First Edition goes to the heart of the matter:

> This book contains almost all the essential material for the biography of a very remarkable man, who died tragically in June, 1954, in the prime of his life and in the middle of research which may still prove to be even more original and important than the finished work which had brought him so much honour and fame.[6]

Irvine stresses the tragedy of Alan Turing's death, a man who perished in his prime, and then proceeds to the matter of the cause of his death:

> Alan Turing's mother, who has assembled and written this record. of his childhood and his mature achievements, believes that his death was

> accidental. The explanation of suicide will never satisfy those who were in close touch with Alan during the last months and days of his life, however much the available evidence may point to it, and in the future the possibility of accident will be considered by those in a better position perhaps to decide the truth.[7]

Neither his mother nor his friends, which would include Lyn Irvine, thought the death was a suicide. The expectation might be that Irvine would continue in this vein, but, instead, she takes another tact: "But even if his death was not chosen by him, he was a very strange man, one who never fitted in anywhere quite successfully."[8] This assessment seems to be based on the idea that fitting in is a good thing. Irvine quotes Shaun Wylie and Sir Geoffrey Jefferson to corroborate her point.[9] All this stresses Alan's indifference to comfort and a certain Spartan nature. Irvine mentions how she introduced Alan to Tolstoy and how "he was totally uninterested in poetry and not particularly sensitive to literature or any of the arts, …."[10] There is an ambivalence in Irvine's portrait of Alan Turing, even more so that the subsequent fuller biography by Sara Turing, and it is in some way an odd choice to introduce a book meant to further Alan's legacy. Here is a positive statement from Irvine: "Alan had recognized himself and his own problems in *War and Peace* and Tolstoy had gained a new reader of a moral stature and complexity and an originality of spirit equal to his own."[11] This is high praise indeed. Then come the next sentences:

> With ninety-nine people out of a hundred Alan protected himself by his off-hand manners and his long silences – silences finally torn up by the shrill stammer and the crowing laugh which told upon the nerves even of his friends. He had a strange way of not meeting the eye, of sidling out of the door with a brusque and off-hand word of thanks.[12]

This may be, from Irvine's point of view, an honest portrait of his strangeness but a curious one to a book by a mother, who admits her son's difference but tries to give a loving and sympathetic portrait of a well-liked and well-respected genius, even of a genius not always understood. Irvine's physical description of Alan, which she continues, is not flattering but worthy of an able novelist, perhaps something out of Charles Dickens. Thus, Irvine oscillates between praise and blame, a kind of epideictic rhetoric and ambivalence. She returns to Sara and her support of her son and her biography: "To those who like myself came to know Alan only in the last ten years of his life, there is the answer to many questions in this short book."[13] Although, unlike Sara, she did not know Alan all his life (only for the last decade), Irvine makes some interesting observations: "He himself found the idea of deceiving others so distasteful that he supposed it equally so to almost everyone."[14] Irvine sets out the nub of what guided Alan Turing:

> Turing quotes his writing to her as a little boy from his preparatory school saying: 'I seem always to want to make things from the thing that is commonest in nature.' This throughout was his ruling principle, and some of the ways that it guided and affected his research are described in this book.[15]

She also relates this principle to Alan's long-distance running. She speaks of his recognition of nature as a child that signalled his "genius." She does not use this word lightly: "There are very few men and women of genius in any century, and of these few some are certain to be overlooked in their own time and possibly in all time."[16] Irvine concludes with a final focus on mother and son: "Sara Turing survives her son owing to the tragic earliness of his death, and with great courage and faith she has taken the opportunity that his death offered and made this source-book for a future biographer."[17] Next Irving amplifies on this courage, and she ends with a sentence that justifies Sarah Turing's biography of Alan Turing and, in some ways, my discussion of this representation and the portraits and self-portraits in my book: "Nothing that science can ever offer is more valuable than the knowledge of how a scientist develops."[18]

The body of the biography begins with family, which is important to Sara Turing as it was for Marie Curie. Sara begins with this family of Norman background arriving in Scotland in 1316. Sara Turing notes that John Robert Turing, "Alan's grandfather, was admitted to Trinity College, Cambridge, in May, 1844, and in the Mathematical Tripos 1848 was classed eleventh among the 'Senior optimes.'"[19] Besides establishing lineage and an old and distinguished family, Sara Turing shows that mathematics had been in the family, although this John Turing became a clergyman. Sara also gives a brief portrait of his son:

> Alan's father, Julius Mathison Turing, was born, 9th November, 1873. On the death of his father, when Julius was ten, the family moved to Bedford: later from Bedford School Julius won a history scholarship to Corpus Christi College, Oxford, and thence passed into the Indian Civil Service, and was posted to the Madras Presidency. He inherited none of his father's mathematical ability,....[20]

Sara does not pretend that either her husband or she understands their son's mathematics. She traces her side of the family back:

> On the maternal side Alan was descended from the Stoneys. According to Burke's Landed Gentry of Ireland the Stoneys are believed to be descended from a Danish family which settled near Kettlewell in Craven in Yorkshire about the ninth century and were known by the name 'de Stanehow,' or 'Stonehow.'[21]

There is an Irish connection in this family: "George and Mary Stoney emigrated to Southern Ireland at the end of the seventeenth century obtaining land under the William and Mary scheme which offered inducements to English Protestants with capital to settle there."[22] Some of the portraits show Sara Turing having an eye for the dramatic as well as a sense of history:

> Thomas George Stoney, J.P., of Kyle Park, Co. Tipperary, Alan's maternal great-grandfather, married in 1829, Anna Henrietta Waller, a member of the family of Wallers, among whom were Sir William Waller (known by Londoners as "William the Conqueror"), a highly skilled General in Cromwell's army, and his first cousin, Hardress Waller. The latter was one of the Regicide Judges; but in 1660 he professed his penitence, adding that he "did appear more to preserve the King upon trial and sentence than any other." His petition for pardon is among the Egerton manuscripts in the British Museum.[23]
>
> This is no ordinary family. Sara also establishes the strain of engineering in her family:
>
> Francis G.M. Stoney and Edward Waller Stoney (my father), became civil engineers. The former, Alan's great-uncle, invented the "Stoney Sluice" used on the Assuan Dam, the Manchester Ship Canal, in the bridge over the Thames at Richmond and at numerous other places the world over; he was also the inventor of the "Titan Cranes."[24]

Sara amplifies this engineering prowess with a portrait of her father:

> Edward Waller Stoney, C.I.E., Alan's maternal grandfather, spent most of his professional life as an engineer of the Madras and Southern Mahratta Railway, of which he later became Chief Engineer. His inventiveness came out in the original methods he devised for the construction of bridges over some of the great Indian rivers, notably the Tangabudra.[25]

Both sides of Alan Turing's family had a connection with India. Sara seems to be establishing more than just a family context but also a genealogy of Alan's character and mind. Like Marie Curie, Sara Turing is facing an unexpected and premature death and turns in part to family to heal her trauma and to bolster or secure the ongoing reputation of the lost loved one. Sara provides another portrait of another relative:

> A distinguished collateral relation of Alan's was Dr. George Johnstone Stoney, F.R.S., president of the Royal Dublin Society. He pre-supposed the existence of the electron and in its hypothetical stage named it and also gave the name to ultra-violet rays. He was known as "Electron Stoney": with all his learning he used to say, "we know so little." He was one of the great who nevertheless retained into old age a childlike simplicity.[26]

Sara Turing shows respect for science and for the curiosity and focus of the scientist. After some other family portraits, Sara comes to her own self-portrait:

> I was Ethel Sara, daughter of Edward Waller Stoney, and was born at Podanur, Madras Presidency, on 18th November, 1881. My education was at the Alexandra School and College, Dublin, and at Cheltenham Ladies' College. Later I attended lectures at the Sorbonne in Paris, before I joined my parents in Madras.[27]

Sara then describes her meeting of Alan's father: "On a voyage home from India via Japan, Canada and the United States of America I met Julius Turing and we were married in Dublin on 1st October, 1907."[28] The chapter on the family ends with her.

The next chapter is on Alan's childhood and early boyhood and begins with the birth of Sara Turing's children:

> Shortly after our marriage my husband and I returned to the Madras Presidency. Our elder son, John Ferrier, was born at my parents' home in Coonoor, Nilgiris, 1st September, 1908. Nearly four years later our second son, Alan Mathison, was born at Warrington Lodge, Warrington Crescent, Maida Vale, London, on 23rd June, 1912.[29]

Born in India, John, as we shall see, writes about Alan, born in England. Very soon, Sara Turing becomes explicit in representing the promise of Alan Turing even as a young child: "My letters to my husband when I was in England in the spring and summer of 1915, round about Alan's third birthday, give some idea of what he was like" and she gives corroboration within the family:

> I was not alone in my opinion when I wrote, 'a very clever child, I should say, with a wonderful memory for new words,' for I reported that his uncle, Herbert Trustram Eve, maintained that he would do great things – this when Alan was nearly three.[30]

Sara provides textual evidence at the time – extracts from letters. She also mentions the times her husband and she were away from their sons. Another of Alan's gifts as a child, Sara Turing represents: "Alan was interested in figures – not with any mathematical association – before he could read and would study the numbers on lamp posts, etc."[31] She gives a detailed account of how Alan as a small child devised a method for remembering how to read numbers left to right and right to left.[32] Sara presents herself as prescient:

> When he was about six years old his quite original comments on or descriptions of things led me to suspect unusual gifts and the likelihood of his becoming an inventor: hence my retention of so many school reports and letters.[33]

She claims that her saving of documents was proof of her suspicion that Alan's early gifts meant he would be an inventor. This assertion builds on her family history full of those gifted in mathematics, engineering and invention. Sara is good at constructing a narrative, a kind of retrospective prospect: she looks back on her looking ahead. She professes her faith in Alan. He did well, and would Sara have written the biography had he not done so well? The reader might well trust Sara, but there is uncertainty as no one can predict the future. One might have a feeling or an intuition or faith, but human fallibility leaves us blind to the future. We can predict, but our words go into possibility and perhaps even probability but not certainty. How well did she know her child as she was away a good deal and Alan went to boarding school and then university? How much is she following how Alan presented himself to her or in his writing or to his friends who gave accounts of him? These are questions I cannot answer, and although her son John later casts doubt on her prescience, he may have his own motives conscious or unconscious. Sara writes of Alan:

> When it came to reading he was not very responsive to teaching. However, finding a copy of *Reading without Tears*, he discovered the principle (as he said later) and taught himself to read in about three weeks. It was characteristic of his lifelong habit of achieving things in his own way and working from first principles.[34]

Interpreting Alan Turing, even as a child, is no easy task: "Despite adverse reports he made an impression on Miss Taylor, the Headmistress. On his leaving the school at about nine years old she said to me, 'I have had clever boys and hard-working boys, but Alan has genius.'"[35] Here, explicitly is the story of a genius, one recognized from an early age.

Besides Alan's independence and to work from first principles, Sara notes another gift of her younger son: "Instances come to mind of instantaneous intuition."[36] She provides a number of examples of these shortcuts. Sometimes she does not name the source explicitly: "One of his Sherborne masters agrees with my surmise that Alan had a sort of 'periscopic' mind which enabled him to see beyond intervening arguments to some conclusion."[37] Sara Turing's surmise, as she sees it, is not "a wild surmise" that John Keats describes in "On First Looking into Chapman's Homer." She has her corroboration by a Sherborne master. Looking beyond, having a wide and comprehensive view, is something Alan Turing possesses in her hypothetical view.[38] Sara quotes Alan's paper "Systems of logic based on ordinals," which discusses the role of intuition and ingenuity in mathematics, which seems to show her attempt to understand him and to craft a narrative that represents Alan Turing as having these qualities from an early age, apparently as predictions.[39] John of Gaunt and some of Shakespeare's characters in his history plays are prophets, who have what I have called retrospective prophecy because Shakespeare and his audience knew the outcome because the War of the Roses was well in the past for them but not for Shakespeare's historical characters. Sara Turing knew that Alan Turing was a gifted mathematician and used a typology of present and past (as did the authors of the New Testament with the Old Testament) to connect a past future with the present, all done in retrospection. I admire the effort Sara put into the making of this portrait of Alan Turing, especially when he was left in the shadows and neglected for years after his death. She set out to make his legacy. That is no small feat. That is no small task. Imagine a mother born in 1881, giving birth to a son in 1912, learning of his death in 1954 and dying in 1976 with the knowledge that there was no apology from the state he served or recognition of the wrong it had done to him. The suffering mother is one part of the author. Whatever vanity, wishful thinking, rationalization and all too human traits are at play here, a mother losing a child, perhaps even the coldest of mothers experiencing such a loss, is painful: one cannot leave out the suffering even if physical (bearing a child and giving birth is very physical) of a mother. Sara's self-portrait, even if disputed by her son John, is of an engaged, concerned and understanding mother. One way or another, I imagine she suffered. At the very least, she tried to heal or perhaps redeem herself and the loss of her son through the writing of this book; when with Lear, she could have said: "This tempest will not give me leave to ponder / On things would hurt me more."[40] Writing could be healing, but it could also be a pondering that could hurt more. Family honour was also something that concerned Charles Darwin and Marie Curie.

Sara Turing does not shy away from challenges, including changes owing to absence:

> When I came home in the summer of 1921 I found Alan very much changed. From having always been extremely vivacious – even mercurial – making friends with everyone, he had become unsociable and dreamy. He had greatly missed my husband and myself and his brother John, who was away at a preparatory school.[41]

Separation is a fact of family life and the life of Alan Turing. She speaks about Alan joining his brother, John, at school in January 1922, and Sara says that John "was somewhat disgraced by his younger brother spending most of his spare time his first term making paper boats, possibly as a form of escapism, since boarding-school life deprived him of his usual occupations."[42] The mother intimates that Alan disgraced John, and the family relations then and after among and between them become an issue with John's account only published in 2012, long after the death of his mother and as part of a second edition of his mother's biography of his brother as part of the centenary of Alan's birth. She elaborates on the painful times:

> The autumn saw the boys back at school – John back to Marlborough where he had gone the previous term and Alan back to Hazelhurst; we returned to India, but he hated these partings as much as we did and we were left with the painful memory of his rushing down the school drive with arms flung wide in pursuit of our vanishing taxi.[43]

This scene of pain and separation for Alan and, in Sara's view, for his parents – for her husband and her – cannot have been easy to write. At some length, she recounts Alan at school, his interests, his sport and then describes a new separation between parents and children: "In 1924 my husband resigned from the Indian Civil Service and we spent the winter and spring of the following three years at Dinard in France."[44] Still, Sara never wavers from a central theme: "Alan and I often talked about great scientists, particularly Pasteur whose portrait was on the French stamps, and I set before him the possibility of himself making some great contribution to science."[45] This is a summary of conversations, and the reader, as a reader must, considers whether the narrative or argument is reliable as one does with any memory or account about the past or any text.

The next chapter tells of Alan Turning's time at Sherborne School. Sara describes the ups and downs of Alan's experience, including the observation: "Nevertheless he carried off the Plumptre Prize for mathematics and did a remarkable examination paper which won from Mr. (now Colonel) Randolph the tribute, 'A mathematician I think.'"[46] Amidst the narrative, the predictions continue: "Mr. Nowell Smith resigned at the end of 1927. On his retirement he wrote to me describing Alan as 'out of the common run' and as a boy of 'marked originality.'"[47] In the portrait of the young genius, Sara does not mind showing her own limitations in science: "In the Christmas holidays of 1927, Alan, being then fifteen and a half wrote for me a précis of one of Einstein's books on Relativity in order to help me to understand the subject."[48] It is hard for Alan to speak for himself, and he does not seem to have written about himself, and some documents may have been damaged or lost:

> Alan kept a private locked diary. Unfortunately one boy out of mischief or from some other motive forced the lock and, I believe, irreparably damaged the book, in which was probably entered mathematical research. This piece of wantonness has deprived us of valuable records from which his early development might have been traced. The loss very much distressed Alan.[49]

Another distress that Sara recounts is another kind of loss:

> In the close friendship and memory of his friend, Christopher Morcom, Alan found special inspiration, but this boy who had just won a Natural Science Scholarship to Trinity College (Cambridge) died in February 1930. How deep was his attachment is revealed to me in a letter at the time. "I feel that I shall meet Morcom again somewhere and that there will be some work for us to do together as I believed there was for us to do here. Now that I am left to do it alone I must not let him down, but put as much energy into it, if not as much interest, as if he were still here.
>
> ["][50]

Sara elaborates on the letters kept surrounding this death, Alan's vacation with Colonel and Mrs. Morcom and Alan's winning of the prize in natural science that the Morcoms awarded in the name of their son. Sara quotes A.J.P. Andrews, a science master, who twenty-four years after Alan Turing left Sherborne, writes: "I have always thought Alan and his friend Christopher Morcom were the two most brilliant boys I have ever taught."[51] Despite recognizing Alan Turning's brilliance, the Headmaster Mr. O'Hanlon wrote Sara Turing in 1954:

> But the picture I keep and treasure is of a somewhat untidy boy arriving during the Railway Strike, after making his way on a bicycle from Southampton via the best hotel in Blandford, and reporting, 'I am Turing.' It was a good start.[52]

This is just a sliver of Sara's narrative and quotations about Alan. She also makes the home sound like a good-natured place in which Alan was the expert on science.[53]

The subsequent chapter is about Alan Turing at Cambridge, which begins thus:

> Being a scholar born, Alan found himself thoroughly in his element when he went up to King's as a mathematical scholar in October 1931. The balance of freedom and discipline just suited him. He at once took up rowing, being in the College trial eights in 1931, 1933 and 1934.[54]

Among the details and testimonies of others, Sara Turing mentions that Alan was best man to his brother at his wedding.[55] She notes that Alan "was fully occupied with writing his thesis on the 'Gaussian Error Function,' which won him his election into a Fellowship at King's in March 1935 – before he was twenty-three."[56] Just as Erasmus Darwin was not there to respond to Charles Darwin's biography of him or Pierre Curie to react to Marie Curie's biography of him, Alan Turing did not live to comment on Sara Turing's biography of him. She seems to want the reader to see that Alan gave his approval: "When I proposed putting together a few notes of his early days for use by a biographer in time to come, he gave an approving grunt."[57] Although the sentence is ambiguous, it appears to mean that she, not he, would gather notes for his future biography. Sara also interprets his grunt as approving her proposal. This is not an explicit authorization but a scene to establish the legitimacy of her biographical enterprise. She interprets this grunt as part of his modesty, while he has a sense of his work:

> Yet this attitude was quite dissociated from any personal conceit; for instance, he was quite taken aback, when he was about to give a lecture in London, to receive a message of greeting from Bertrand Russell, as he hardly thought he would come within the ken of so eminent a mathematician and philosopher.[58]

Thus, Alan has confidence without conceit. Here, there is some overlap of figures I discuss in my examination of representation and self-representation of scientists, for instance, Bertrand Russell and Alan Turing, the Darwins and the Grays. On Alan Turing's work, Sara Turing skips ahead from the 1930s and quotes from *The Times* obituary of 16 June 1954, Robin Gandy's comments in *Nature*, 18 September 1954 and M. H. A. Newman's memoir for the Royal Society in November 1955.[59] Here, she defers to experts and quotes from them, allowing for their positive assessments to accumulate. Sara discusses her walks with her son as he explained his research to her, and she mentions Alan's mathematical papers.

The next chapter is about the graduate college at Princeton. Sara quotes from a letter Alan wrote a week after arriving there:

> The mathematics department here comes fully up to expectations. There is a great number of the most distinguished mathematicians here – J. v. Neumann, Weyl, Courant, Hardy, Einstein, Lefschetz, as well as lots of small fry. Unfortunately there are not nearly so many logic people as last year.[60]

She outlines what Alan did at work and for leisure at Princeton and what he liked and did not like and quotes a letter from Dean Eisenhart to her near the end of Alan's first year at the Graduate College: "We have enjoyed having him here very much, not only for his personal qualifications but for the impression he has made on our men in mathematics."[61] Throughout the biography, Sara Turing stresses and includes quotations that stress the positive personal qualities and mathematical abilities of her son. Sara reports that after the vacation in England during the summer of 1937, Alan went back to Princeton for a second year and worked on his PhD dissertation, "Systems of Logic based on Ordinals," under the direction of Alonzo Church, receiving his degree in May 1938. Furthermore, Sara quotes from the memoir published in an annual report of King's College, Cambridge, in November 1954, emphasizing the importance of the work on ordinal logics at Princeton, the paper on the word-problem and another on computable numbers.[62] Alan consults but makes his own decisions: "My husband advised Alan to find a job in America. Alan himself, though not very keen, made tentative enquiries, but eventually refused the offer of a post as von Neumann's assistant."[63] Instead, he returned to King's, which renewed his fellowship.

Sara chooses something different for the next chapter, which is "Some Characteristics." She describes his physical appearance, describing him as youthful or sometimes childlike, whose dress could be "slovenly" and his hair too long.[64] Moreover, she says that Alan "could be abstracted and dreamy" and notes that "'vivid, lovable, modest,' were the epithets constantly employed by his friends in writing of him."[65] She tells of his generosity and kindness, not having his "head was not always in the clouds" but he "took a real interest in his fellow creatures."[66] To corroborate, she quotes Sir Geoffrey Jefferson:

> He was so unversed in worldly ways, so childlike it seems to me, so unconventional, so non-conformist to the general pattern … so very absentminded. His genius flared because he had never quite grown up. He was, I suppose, a sort of scientific Shelley.[67]

Sara interprets Alan's character further, including his stammer, and his generosity and choice of gifts, his rapport with children.[68]

Sara Turing begins the next chapter, "War Work in the Foreign Office," with a description of the context:

> We had given up our home in Guildford in March 1939, and being mostly on the move until war broke out we saw little of Alan: during the next six years he could spend only snatches of leave with us, for immediately on the declaration of war he was taken on as a temporary Civil Servant in the Foreign Office, in the Department of Communications.[69]

Sara explains that his work was secret and focuses on what he could tell her, what she calls "Alanesque things," and mentions that his nickname was "the Prof" or "Prof" at Bletchley Park.[70] She returns to the reports of others and the theme of Alan Turing being wrapped up in theories with a disregard to hair, clothes and conventions. Alan's wandering around fields causing suspicion, his knitting of gloves and his bicycling with a gas mask on as part of a series of amusing stories. She alludes to Professor Newman's memoir of Alan from the Royal Society and quotes A.C. Chamberlain on Alan's indifference to scorn, his determination and method using theory and observation.[71] Donald Michie is someone Sara quotes to confirm her own views of Alan's genius, his great intellect, his gusto for topics, his heart, his childlikeness and goodness, which made "him a very vulnerable person in a world so largely populated by self-seekers."[72] Sara accumulates the testimonies including one fellow worker at Bletchley (perhaps unnamed because of the secret nature of the work) about Alan's "brain power" and impressive "championing of the underdog and his willingness to help others who had problems which to him were just simplicity itself."[73] Whatever differences of opinion or reservations others might have had with Sara Turing's portrait of Alan Turing, I would note the care she took to consult and read others who knew and worked with Alan and her portrait of a generous, kind, vulnerable and brilliant man – her son – in the wake of his death when the authorities seemed to have chosen to punish him rather than thank him.

In her amusing anecdotes meant to humanize Alan, Sara describes his crowded and difficult voyage in September 1942 to the United States and admission there. She also shows how resourceful he was in solving problems even outside of mathematics:

> The voyage to America of course had considerable hazards and for me, some anxiety, but he circumvented the censor by sending me from New York a cable of good wishes for my birthday, which assured me of his safe arrival.[74]

Sometimes Sara is left to speculation: "He remained in America until March 1943. He seems to have taken the opportunity to visit Princeton and probably saw something of the progress of computing machinery in the States."[75] She quotes an unnamed friend from Bletchley who commented on Alan's originality, his view of authority that it was based on "a better grasp of the subject involved than anyone else," his assumption that people would "listen to reason" and his "practical weakness ... that he wouldn't suffer fools or humbugs as gladly as one sometimes has to."[76] Another source, Dr. I. J. Good, whom Sara quotes, says that "Prof" could offend administrators and directed his humour at authority.[77] After Bletchley, Alan had a new assignment, and Sara then moves to a quotation from A.C. Chamberlain, who comments on Alan's "character" and happy independence, adding that "Alan was easily the most brilliant man I have met."[78] These are some of the testimonies, something Marie Curie does to honour Pierre Curie in death, even if there are fewer to end her account. Sara stresses how Alan could simplify a problem and was simple and unassuming. She also mentions his running and tells how he won the regimental mile race. She quotes the commanding officer:

> We rather thought it might be a leg-pull, but on the Day 'The Prof.,' by which name Turing was also known, came in a very easy first. He left us to work on a new Electronic Brain, and we saw him no more, but all were very sorry to part with such a pleasant messmate who was liked and respected by all ranks.[79]

Part of Sara's narrative depends on her own relation to Alan and her own experience, which she admits was sometimes lacking owing to circumstances: "Occasionally during the war Alan came to stay with me for, a week or so in Guildford, but owing to the secrecy of his work there is little to report."[80] Alan does not seem to have been too worried about titles and honours:

> For his war service he was made an Officer of the British Empire, as he was one of a team whose joint work was an important factor in our winning the war. Some friends dining with him some years later wanted a nail or screw for something. Alan produced a box in which were odds and ends, nails, etc., and his O.B.E. medal all mixed up together, much to the amusement of his guests.[81]

Sara says that Alan chose the practical over the conventional, which led to tales that Alan and others dismissed as legendary.

Chapter 8, "At the National Physical Laboratory, Teddington," opens with the offer of a Cambridge University lectureship after he left the Foreign Office but also with his declining of it "since his attention was becoming focussed on computing machinery."[82] She quotes the Director of National Physical Laboratory, Sir Charles Darwin, from a broadcast of November 1946:

> The project has been picturesquely called the Electronic Brain. For a long time mathematicians have been occupied in getting better logical foundations for their subject, and in this field, about twelve years ago, a young Cambridge mathematician, by name Turing, wrote a paper which appeared in one of the mathematical journals, in which he worked out by strict logical principles how far a machine could be imagined which would imitate processes of thought.[83]

Sara represents the recognition Alan was receiving, by a Darwin no less. Sir Charles Darwin (1887–1962) – the son of Sir George Darwin, Plumian Professor of Astronomy at Cambridge – was an astronomer and geophysicist, and the grandson of Charles Darwin. I can only imagine a conversation between this Charles Darwin and Alan Turing as E. C. Bullard's obituary says: "Darwin was a man of great intelligence and charm with a most engaging habit of deliberately saying the wrong thing when he thought it would stimulate a discussion" and refers to missing his "cheerful pessimism." Cambridge was full of brilliant "eccentrics," including Sir Charles Darwin and Alan Turing.[84] In a biographical memoir from the Royal Society of the death of Sir Charles, states: "The late A. M. Turing, F.R.S., a brilliant young mathematician on the staff of the N.P.L., was one of the pioneers in the logical design of such computers."[85] In an obituary of Turing the Royal Society, M. H. A. Newman begins with a statement that shows high esteem and for him that takes up a theme that Sara Turing repeats: "The sudden death of ALAN TURING on 7 June 1954 deprived mathematics and science of a great original mind at the height of its power."[86] Moreover, Sara Turing mentions the caricatures in the newspapers, remarking wryly: "Some years later Alan remarked that the daily papers were many years ahead of him, opening even *his* eyes in wonder, so far did they outstrip him in their forecasts."[87] Sara says that Alan ran eighteen miles from Hampton to Guildford and fourteen miles from Teddington to Dollis Hill rather than take public transport with many transfers and walked after midnight – when the last train left – sixteen miles between Surbiton and Guildford.[88] Alan's father died in August 1947. Alan went on sabbatical to King's, where he was a fellow. Sara says she gleans information of his sabbatical from Professor Pigou and Mr. Peter Matthews and Dr. J. W. S. Pringle. She quotes Professor Pigou:

> Though primarily a mathematician, he was interested in many other things as well and would gallantly attend lectures on psychology and

> physiology at an age when most of us were no longer capable of sitting on a hard bench listening to someone else talking.[89]

A colleague attests to Alan Turing's wide interests and deep curiosity. Sara concludes about Alan: "Not only has his dream come true, but his ideas have profoundly influenced computer designers throughout the world."[90] Alan's legacy Sara sets out as she concludes the chapter.

"Work with the Manchester Automatic Digital Machine," Chapter 9, opens with a summary of a new transition: "On resignation from the Scientific Civil Service Alan accepted a Readership at Manchester University and was appointed Assistant Director of 'Madam,' the Manchester Automatic Digital Machine designed mainly by Professor F.C. Williams, F.R.S., and Dr. T. Kilburn."[91] She quotes Denis Williams on Alan, which includes this observation: "In intellectual, as in other matters, it was essential to him that everything should ring true."[92] Sara adds more about the interaction of Alan and her about this time: "Sometime round about 1944 he had talked to me about his plans for the construction of a universal computer and of the service such a machine might render to psychology in the study of the human brain."[93] Cybernetics is Sara's next topic, and she mentions Professor Norbert Wiener (1894–1964), who visited England in 1949, who said that Alan's work "represented an original combination of the modern stream of mathematical logic with the theory and practice of control and communication apparatus."[94] Wiener himself is an interesting scientist, someone Anders Lindquist and I have discussed on many occasions, and I have read Weiner's manuscript written as a schoolboy, "The Theory of Ignorance"(1906), a precocious essay. This handwritten essay begins the definition of "general ignorance" as the impossibility of humans being certain or sure of anything.[95] In logic and mathematics, figures like Russell, Wiener and Turing overlap. Sara quotes others on Alan's work, including Professor M.H.A. Newman's view comments on Alan's "Solvable and Unsolvable Problems": "It was a remarkable feat to give a proof of the fundamental 'unsolvability' theorem understandable by readers with no previous knowledge but willing to think rather hard."[96] For Sara, an honour Alan is given connects him to the family context she set out in the first chapter:

> His election as a Fellow of the Royal Society took place in the spring of 1951; his proposer was Professor M.H.A.Newman, F.R.S., and seconder, Bertrand Russell, O.M., F.R.S. He thus became the fourth member of my family to be elected a Fellow of the Royal Society.[97]

It seems that Russell also thought highly of Turing. Sara also reports on other developments: "In 1952 he resigned his fellowship at King's College, Cambridge, where he had been a Fellow for seventeen years. Next year he gave a lecture on computing machinery at his old school, Sherborne."[98] This was his present facing the past.

Chapter 6, "Broadcasts and Intelligent Machinery," opens by going back to 1951, reminding us that sometimes Sara follows themes and can jump back and forth in time:

> On 15th May, 1951, on the Third Programme, Alan gave his first broadcast, which was one of a series with the general title 'Automatic Calculating Machines.' His lecture, the second in the series, bore the sub-title, 'Can Digital Computers Think?[99]

Sara quotes Professor Sir Geoffrey Jefferson from a letter that "spoke of 'Alan in whom the lamp of genius burned so bright... he had real genius, it shone from him.'"[100] She shifts back and forth in time in her shaping of the story of the genius of Alan Turing.

In "Morphogenesis," Chapter 12, Sara Turing explains the meaning of the word she uses as a title as one that describes "the processes involved in the appearance and development of living structures."[101] She is willing to admit a lack of knowledge: "How long Alan had been ruminating on his chemical theory of the growth of living things, particularly the mathematical aspect of it, we do not know."[102] This interest was something Alan Turing developed in publications that includes his "monograph on morphogenesis was contributed in November 1951 to the *Philosophical Transactions of the Royal Society*...."[103] Sara wonders whether Alan's theory of living tissue might apply to cancer.

Chapter 13, "Relaxation," sets out another theme, a reminder that Sara's biography is often more thematic than strictly chronological. She picks up on earlier threads from his childhood and youth: "Alan was a very hard worker, but when he relaxed he did so thoroughly, running, walking or cultivating his garden."[104] Sara also continues to stress the bond between mother and son and her attempt to understand Alan:

> On his short visits to Guildford in the vacations he always put in a good deal of work, yet he found leisure for walks with me and long discussions. Despite the great mental disparity between us we never lacked topics of conversation.[105]

She shows her humility and realism in representing her relationship with her gifted son. In discussing Alan's sailing, Sara creates a typology between the child and the man: "Sometimes he took a short holiday sailing at Itchenor with my husband or alone. Once on a trip alone to the Isle of Wight he narrowly escaped being run down by a large vessel" and she adds: "As a boy he enjoyed riding; so, what with rowing, mountain climbing, hockey, tennis, cycling and long distance running his range of exercise was varied, while for indoor games there was his interest in chess and 'Go.'"[106] Alan Turing liked to play games, enjoyed sport, was active outdoors. Sara shifts to Alan's reading, his preference for encyclopaedias and scientific work as a boy and reading some fiction as a teen: "He had a particular fondness for *The Pickwick Papers*, George Borrow's books and Samuel Butler's *Erewhon*. This last possibly set him to think about the construction of an actual intelligent machine."[107] Sara lists the works Alan liked such as novels by Jane Austen, William Makepeace Thackeray and Anthony Trollope while enjoying more recent novels by Dorothy Sayers and John Galsworthy. Poetry was another matter: "Poetry, with the exception of Shakespeare's, meant nothing to him: he liked people to state plainly what they had to say. The poet might reply: 'What about your symbols?'"[108] Sara imagines a dialogue between Alan and the poet. She also says that Alan enjoyed Tolstoy – *Anna Karenina* and *War and Peace* – as well as plays broadcast. Once more, Sara returns to Alan's athletics and his winning mile and three-mile races. He even ran a race just before Christmas in 1946, the same day he sailed from Southampton for the United States. Anyone knowing the geography of Surrey and nearby would know that the logistics are impressive. An injury kept him from the Olympic trials, and Sara quotes Professor Pigou, who concludes:

> Most unfortunately he displaced something in his hip, and, in spite of the efforts of bone-setters, had to give up the idea of long distance racing. It would have been a feather in the cap of King's College to have had one of its Fellows coming in first in an Olympic marathon.[109]

Alan still ran well but could not race as seriously as he would have liked.[110]

My next chapter concentrates on the end of Sara's biography of Alan as well as tributes to him and some of what other biographers, Alan Hodges and Martin Davis, both mathematicians, say about him and sometimes about the representations of Alan by her and her son John. These responses can suggest something that Turing and none of us can control, what others say about us, especially in death.

Notes

1 Anders Lindquist to Jonathan Hart, email from KTH to Harvard, 19 November 2024. My thanks to Anders Lindquist for granting me permission to reproduce his words here and for his friendship and brilliance in discussions of a wide range of subjects, scientific, social, cultural, historical and more. See Norbert Wiener, *Cybernetics: Or Control and Communication in the Animal and the Machine* (Cambridge: MIT Press, 1948); Arthur Samuel, "Some Studies in Machine Learning Using the Game of Checkers," in *IBM Journal of Research and Development* 3 (3) (July 1959): 210–229.

2 John Searle, "The Chinese Room," in *The MIT Encyclopedia of the Cognitive Sciences*, edited by R.A. Wilson and F. Keil (Cambridge: MIT Press, 1999), quoted in David Cole, "The Chinese Room Argument," *The Stanford Encyclopedia of Philosophy*, edited by Edward N. Zalta & Uri Nodelman (Winter 2024 Edition), https://plato.stanford.edu/archives/win2024/entries/chinese-room/.

3 John Searle, "Why Dualism (and Materialism) Fail to Account for Consciousness," in *Questioning Nineteenth Century Assumptions about Knowledge*, edited by Richard E. Lee (III: Dualism) (New York: SUNY Press. 2010), quoted in Cole, "The Chinese Room Argument."

4 My discussion of the Chinese Room Argument in regards to Searle and Turing is indebted to the longer and more detailed analysis in Cole, "The Chinese Room Argument." See also David Cole, "Thought and Thought Experiments," *Philosophical Studies* 45 (1984): 431–444; John Searle, "Minds, Brains and Programs," *Behavioral and Brain Sciences* 3 (1980): 417–457; John Searle, *Minds, Brains and Science* (Cambridge, MA: Harvard University Press, 1984); Alan Turing, *Intelligent Machinery: A Report* (London: National Physical Laboratory, 1948); "Computing Machinery and Intelligence," *Mind* 59 (1950): 433–460.

5 Sara Turing, "Preface," in Sara Turing, *Alan M. Turing Centenary Edition with a Foreword by Martin Davis and an Afterword by John Turing* (Cambridge: Cambridge University Press, 2012), 4. First edition published by W. Heffer & Sons, Ltd 1959; Foreword to the First Edition © L. Irvine 1959, 2012 Foreword to the Second Edition © M. Davis 2012; Afterword © J. Turing 2012, xviii.

6 Lyn Irvine, "Foreword," in Sarah Turing, *Alan M. Turing Centenary Edition with a Foreword by Martin Davis and an Afterword by John Turing* (Cambridge: Cambridge University Press, 2012), xix.

7 Irvine, "Foreword," xix.

8 Irvine, "Foreword," xix.

9 Irvine, "Foreword," xix–xx.

10 Irvine, "Foreword," xxi.

11 Irvine, "Foreword," xxi.

12 Irvine, "Foreword," xxi.

13 Irvine, "Foreword," xxii.

14 Irvine, "Foreword," xxii.

15 Irvine, "Foreword," xxiii.

16 Irvine, "Foreword," xxiii.

17 Irvine, "Foreword," xxiv.

18 Irvine, "Foreword," xxiv.

19 Sarah Turing, *Alan M. Turing Centenary Edition with a Foreword by Martin Davis and an Afterword by John Turing* (Cambridge: Cambridge University Press, 2012), 4.

20 Turing, *Alan M. Turing*, 4.

21 Turing, *Alan M. Turing*, 4.

22 Turing, *Alan M. Turing*, 5.

23 Turing, *Alan M. Turing*, 5.

24 Turing, *Alan M. Turing*, 6.

25 Turing, *Alan M. Turing*, 6–7.

26 Turing, *Alan M. Turing*, 7.

27 Turing, *Alan M. Turing*, 8.

28 Turing, *Alan M. Turing*, 8.
29 Turing, *Alan M. Turing*, 9.
30 Turing, *Alan M. Turing*, 9.
31 Turing, *Alan M. Turing*, 10.
32 Turing, *Alan M. Turing*, 10–11.
33 Turing, *Alan M. Turing*, 11.
34 Turing, *Alan M. Turing*, 12.
35 Turing, *Alan M. Turing*, 12.
36 Turing, *Alan M. Turing*, 12.
37 Turing, *Alan M. Turing*, 13.
38 *Oxford English Dictionary*, "periscopic (*adj.*)," July 2023.
39 Turing, *Alan M. Turing*, 13.
40 William Shakespeare, III.iv. 24–25, "King Lear," *The Riverside Shakespeare*, edited by G. Blakemore Evans (Boston: Houghton Mifflin, 1974), 1276.
41 Turing, *Alan M. Turing*, 16.
42 Turing, *Alan M. Turing*, 16.
43 Turing, *Alan M. Turing*, 17.
44 Turing, *Alan M. Turing*, 21.
45 Turing, *Alan M. Turing*, 22.
46 Turing, *Alan M. Turing*, 26.
47 Turing, *Alan M. Turing*, 29.
48 Turing, *Alan M. Turing*, 29.
49 Turing, *Alan M. Turing*, 34.
50 Turing, *Alan M. Turing*, 35.
51 Turing, *Alan M. Turing*, 36.
52 Turing, *Alan M. Turing*, 38.
53 Turing, *Alan M. Turing*, 39.
54 Turing, *Alan M. Turing*, 40.
55 Turing, *Alan M. Turing*, 44.
56 Turing, *Alan M. Turing*, 44.
57 Turing, *Alan M. Turing*, 45.
58 Turing, *Alan M. Turing*, 45.
59 Turing, *Alan M. Turing*, 46–47.
60 Turing, *Alan M. Turing*, 51.
61 Turing, *Alan M. Turing*, 53.
62 Turing, *Alan M. Turing*, 54.
63 Turing, *Alan M. Turing*, 54.
64 Turing, *Alan M. Turing*, 56.
65 Turing, *Alan M. Turing*, 57.
66 Turing, *Alan M. Turing*, 57.
67 Turing, *Alan M. Turing*, 58.
68 Turing, *Alan M. Turing*, 59–66.
69 Turing, *Alan M. Turing*, 67.
70 Turing, *Alan M. Turing*, 67.
71 Turing, *Alan M. Turing*, 68–69.
72 Turing, *Alan M. Turing*, 70.
73 Turing, *Alan M. Turing*, 71.
74 Turing, *Alan M. Turing*, 72.
75 Turing, *Alan M. Turing*, 72.
76 Turing, *Alan M. Turing*, 73.
77 Turing, *Alan M. Turing*, 73.

78 Turing, *Alan M. Turing*, 74.
79 Turing, *Alan M. Turing*, 75.
80 Turing, *Alan M. Turing*, 75.
81 Turing, *Alan M. Turing*, 76.
82 Turing, *Alan M. Turing*, 77.
83 Turing, *Alan M. Turing*, 78.
84 See Edward Crisp Bullard, "Charles Galton Darwin (obituary)," *Quarterly Journal of the Royal Astronomical Society* 4 (9) (1963): 316–317. See also the Times obituary of George Darwin at https://mathshistory.st-andrews.ac.uk/TimesObituaries/Darwin/.
85 George Paget Thomson, "Obituary 30 NOV 1963 Charles Galton Darwin, 1887–1962," *Biographical Memoirs of Fellows of the Royal Society* 9 (1963): 78, see 69–85, https://doi.org/10.1098/rsbm.1963.0004
86 Maxwell Herman Alexander Newman, "OBITUARY| 30 NOV 1955 Alan Mathison Turing, 1912–1954," *Biographical Memoirs of Fellows of the Royal Society* (1) (1955): 253, see 253–263, *https://doi.org/10.1098/rsbm.1955.0019.*
87 Turing, *Alan M. Turing*, 79.
88 Turing, *Alan M. Turing*, 84.
89 Turing, *Alan M. Turing*, 85.
90 Turing, *Alan M. Turing*, 87.
91 Turing, *Alan M. Turing*, 88.
92 Turing, *Alan M. Turing*, 91.
93 Turing, *Alan M. Turing*, 92.
94 Turing, *Alan M. Turing*, 92.
95 Norbert Wiener, "The Theory of Ignorance," 1906, MIT Libraries. The title inside the manuscript, "The Theory of General Ignorance," is different from the one on the cover.
96 Turing, *Alan M. Turing*, 98.
97 Turing, *Alan M. Turing*, 98–99.
98 Turing, *Alan M. Turing*, 99.
99 Turing, *Alan M. Turing*, 100.
100 Turing, *Alan M. Turing*, 101.
101 Turing, *Alan M. Turing*, 102.
102 Turing, *Alan M. Turing*, 102.
103 Turing, *Alan M. Turing*, 102.
104 Turing, *Alan M. Turing*, 106.
105 Turing, *Alan M. Turing*, 106.
106 Turing, *Alan M. Turing*, 107.
107 Turing, *Alan M. Turing*, 108.
108 Turing, *Alan M. Turing*, 108.
109 Turing, *Alan M. Turing*, 111.
110 Turing, *Alan M. Turing*, 112.

11 The Last Days and Death

Alan Turing and the Representation of Him

How Sara represents the last days and the death of her son and the tributes, interpretations and biographical accounts of others is the focus here. I begin with the final parts of Sara's biography. Like Pierre Curie, Alan did not leave a sustained memoir or autobiography in print or even in manuscript.

In Chapter 13 of *Alan M. Turing, Last Days and Some Tributes*, Sara Turing approaches the denouement, death and responses. Sara talks about Alan's house and the room he turned into a laboratory and quotes his collaborator:

> Dr. Robin Gandy has given a full account of their joint activities over the week-end a few days before Alan's death. They were busy with the preparation of a non-poisonous weed-killer and sink-cleaner. 'There was nothing sinister about these experiments, which were just an example of Alan's liking to make things himself.[1]

The following is a paragraph that Sara would have had a hard time writing:

> On the 8th June, 1954, Alan's housekeeper found him dead in his bed; death was due to poisoning by potassium cyanide, and it presumably took place on the night of the 7th June. The verdict at the inquest was that the poison was self-administered, while the balance of his mind was disturbed. No poison was found in his bedroom. There was just a partly eaten apple on the table by his bed, for, as a rule, he used to eat an apple at night.[2]

Sara tries to give objective accounts and quotes Dr. Gandy once more and says that many of Alan's friends thought "his death was caused by some unaccountable misadventure. Besides, his inadvertence alone had always involved the risk of an accident."[3] Testimony is important to Sara Turing, and I give a few excerpts of the examples she provides. Sara quotes

DOI: 10.4324/9781003651871-12

Dr. Milner-White, once Dean at King's College Chapel, who expresses his affection and respect for Alan, his brilliance and modest character, "'giving always the feeling that he saw behind many veils which we could not pierce.'"[4] Moreover, Sara quotes Mr. O'Hanlon, his housemaster at Sherborne: "'For those who knew him here, the memory is of an even-tempered, lovable character with an impish sense of humour and a modesty proof against all achievement.'"[5] And she gives her own interpretation of her son that is the reason for her biographical work: "It is my hope that this short biography may serve to irradiate and perpetuate his memory. It is offered in gratitude for a richly endowed life, zealously devoted to the pursuit of knowledge and truth."[6]

Part Two of the biography includes more testimony on Alan's work, which is germane but involves some repetition, so I end here and move to Alan's brother, John, who wrote his account in the 1960s and 1970s and perhaps beyond, and it was published in 2012.

If anything, John Turing is more conflicted or ambivalent about Alan Turing than Sara Turing and Lyn Irvine are about him. As I am concentrating most on Sara's account, I shall mention a few observations of John's that relate to her biography of Alan. John's manuscript, published in the centenary edition as *My Brother Alan*, begins:

> My mother has written a biography of my brother Alan. It was certainly a tour de force on her part to write it, as she did, in her seventies. It has rightly been praised, by others better qualified to judge than I am, as a revealing study of a son who became, undoubtedly, a mathematical genius.[7]

This opening is direct and supportive, but what follows begins with qualification, ambivalence and contradiction. He shifts from praise to blame, support to criticism: "I start off, therefore, at a disadvantage. Much of the ground has already been covered, and certainly, so far as Alan's childhood is concerned, far better and in more detail than I could have done it myself."[8] The word "disadvantage" is understandable but seems slanted. John is responding and supplementing, so after praising his mother's effort to provide a biography of Alan, he speaks as though her work and that of others, despite being better and more detailed than he could have done, leaves him disadvantaged in his view. He continues: "But it is sufficiently obvious to any discerning reader of that book that a false note has been struck somewhere: could Alan have been quite that paragon of virtue that my mother describes?"[9] John realizes that his demythologizing of his brother has its hazards:

> Yet, if an elder brother ventures to suggest the contrary, it can easily be suggested that he is jealous and sour. This is a risk I must accept.

> My only concern is to put the record straight, however hazardous the enterprise.[10]

There does seem to be some sourness, but I shall focus on what John deems as setting the record straight. It is important to have another point of view and especially within the family. There are many interpretations of Alan Turing in the wake of Sara's biography, and even John's account waited for publication in the centenary edition.

John tells the story of Alan's birth from his point of view and discusses their childhood in a reasonable way, including parents and children being separated because of service to the British Empire. John does not blame anyone for this and even says his mother suffered most from the separation.[11] John thinks that Alan's homosexuality arose from his time of being left at the Wards during these separations. Oddly enough, Sara's view of homosexuality is closer to the one that prevailed after the 1960s and 1970s in the West, although there is still a tinge of misunderstanding. John reports that Sara made no reference to Alan's homosexuality in her book, and he colours her response based on John's conversation with her: "her reaction was much what one might expect if a specialist had informed her that her son was colour blind or had an incurable obsession with spiders."[12] John says he does not like his brother being made an early crusader in "the gay crusade."[13] This is setting the record straight about John's own view of homosexuality and makes, perhaps inadvertently, Sara look more tolerant in this regard than John is. John says that Alan was not a misogynist, and John gets into a discussion of a "safe" female, which may reveal more about him than it does about Alan.[14] The description of the Ward household is of interest. For instance, John observes:

> Let the psychiatrists make of it what they will; no doubt they could have a field day of it. I just think, myself, that if Alan had been blessed with, let us say, an understanding and loving sister a year or two older, or even younger, than himself, instead of a sardonic elder brother and the tiresome Joan Ward, he might have turned out differently.[15]

John's self-representation is as ironic and sardonic as he deems himself. John sees that he might not have always had a positive effect on his younger brother, Alan.

Another aspect of setting the record straight is the matter of genius from an early age:

> My mother, perhaps unwittingly, gives the impression in her book that she recognized Alan's genius from the start, and that she sedulously fostered it. If so, she did not give that impression in the family at the time; in fact, quite the contrary.[16]

John also notes that his father generally ignored Alan's eccentricities and speaks for himself:

> As for myself, with the selfishness of youth, and separated by a gap of four years, I did not care what Alan did, and I was content to go my own way, as indeed he was content to go his.[17]

John takes credit for recommending Sherborne for Alan and commends his parents for choosing it over Marlborough, "which all but crushed me and would certainly have crushed him."[18] In addition to irony, John also uses self-deprecation. John represents his mother differently from the way she represents herself in the biography of Alan:

> The only person in the household who was forever exasperated with Alan, constantly nagging him about his dirty habits, his slovenliness, his clothes and his offhand manners (and much else, most of it with good reason) was my mother. If this was due to some early recognition of his genius, she was certainly doing nothing to foster it by trying to press him into a conventional mould.[19]

He is often hard on Alan and praises the headmaster and a master named Blenkins at Sherborne for their conduct in relation to Alan. The not-so-stellar school reports of Alan are something John goes into in detail as well as observing: "Alan was first class at beating the system. He refused to work at anything except his precious maths and science, but he had an incredible aptitude for examinations, aided by last minute swotting."[20] John contrasts his mother with his father and him: "My mother evidently persuaded herself that Alan's sudden and unexpected blossoming at Sherborne, which had dumbfounded my father and myself, was no more than what she had anticipated all along."[21] John thinks his mother had "blind faith" in Alan.[22] This last assertion seems to contradict his previous assertions about his mother's view of Alan. John's views seem ambivalent and contradictory. After recounting some tales of Alan's oddity (something Sara had done), John tells an amusing story of his mother and he sorted the sheets for Alan's thesis at Cambridge in 1935 (he had gone to King's in 1931 on a mathematics scholarship) and how he was more than twenty minutes in advance at the post office. John wryly observes: "This is my only positive contribution to mathematical thought."[23] John then moves to Alan at Bletchley: "Much has been written about this establishment in recent years, including a great deal of rubbish about Alan," and he recommends Peter Calvocoressi's *Top Secret Ultra,* which mentions Alan "'whom even BP's most brilliant cryptographers put in a class of his own.'"[24] John can be quite positive about Alan. He can even agree with Alan and think him underappreciated:

> As someone rightly remarked, for thus saving the nation from disaster, Alan should have been given an earldom but, in fact, he was awarded an OBE, which, to the amusement of his friends, but quite properly, in my opinion, he kept in a tin box along with such aids to gracious living as screws, nails, nuts and bolts.[25]

When Sara tells about this event, she is less sure of the rightness of the medal in the tin box. Although John wrote this account in the 1960s and 1970s, he obviously wrote some of it in the 1980s as he quotes a letter in *The Times* of 12 September 1981 in which Mr. G.M. Watkins suggests a book should be written on the eccentrics at Bletchley, something John supports, and John puts Alan in "the top league" of that group.[26] John even wishes to set the record straight on Alan's eccentric ways: "My brother's eccentricities have become legendary and some of them have become distorted in the process."[27] For John, a distinction among eccentricity, idiosyncrasy and plumb craziness needs to be made, and he gives examples of these three states in stories about Alan. John recounts Alan modestly volunteering for a cross-country race in an army unit he was attached to for a week. He notes that this practical joke was fair:

> The request was granted; all in the mess looked forward eagerly to the Prof. trailing in well behind the rest. Alan had failed to mention that he was a marathon runner of near-Olympic standard (soon after the war, in August 1947, he came fifth in the Amateur Athletic Association Marathon Championship). So, of course, he arrived back three minutes before anyone else.[28]

John sees this behaviour as unpredictably eccentric because Alan "had never shown the slightest interest in athletics before, yet here he was, at the age of thirty-five, in the front rank of marathon runners and seriously considered for the Wembley Olympics."[29] Referring the reader to Sara's book for Alan's "athletic exploits," John does not explicitly recognize that this was one talented eccentric mentally and physically, not some eccentric whose eccentricity is the be all.[30] Setting the record straight is something that John justifies explicitly as he thinks that his mother would have been fairer to Alan had she not mentioned his virtues only:

> My mother implies that his many eccentricities, divagations from normal behaviour and the rest were some kind of emanation of his genius.

> I do not think so at all. In my view, these things were the result of his insecurity as a child, not only in those early days at the Wards, but later on as his mother nagged and badgered him.[31]

Here and elsewhere, John refers the matter to professionals, here psychologists, elsewhere psychiatrists. John specifies his own motives: "I am concerned only with Alan's behaviour as it affected other people and it was not, in my view, so amusing for those who were at the receiving end."[32] The others turn out to be John and the family and instances of Alan showing up day or night on short notice. Moreover, John notes that Alan seldom wrote letters to relatives as he did to his few mathematical peers worldwide but would send family members a postcard or telegram announcing "Arriving today" – "My mother received loads of telegrams of this sort and they drove her wild, but nobody reading her book would think so."[33] John seems to have ambivalent feelings about his mother and brother. Sara seems to have wanted to help create a legacy that showed respect for Alan, and John appears to have wished for a fuller portrait of Alan as a human. The one is in danger of creating a genius saint, whereas the other might deflect from the great scientist and dwell on the all too human child and man. Both portraits give us, from a familial point of view, an intricate portrait that should be considered in relation to Alan Turing's own writing and that of later biographers, mathematicians, historians and others.

John speaks of Alan not liking "social chat," which he considered "vapid," and says that Alan preferred "a thoroughly disputatious exchange of views" as if he were a brilliant contrarian.[34] John summarizes: "Alan's hatred of 'vapid conversation,' his fear of 'unsafe' women and the value he placed on the importance of time – that is to say, his own – did not make him the most amiable or helpful of guests."[35] This sentence encapsulates important aspects of John's theory or view of Alan. John tells of a sherry party at his own house in which Alan came dressed like a tramp and, bored, left after ten minutes without a word: "Frankly, I wished I could have done the same and I greatly envied him for having achieved a *modus vivendi* in which the feelings of others counted for so little."[36] The shifts in John's narrative, including of tone, make it interesting and dramatic but may also embody his ambivalence to Alan. The next sentence in a new paragraph as the text is published is as follows: "The photographs in my mother's book are excellent and show him exactly as he was."[37] Then John makes some comment about Alan being alluring and was thus made a don at 23 and then adds insult to injury (or is perhaps trying to balance against his own innuendo) by saying that Alan was not alluring "to some people," adding

> He lived, as I suppose is the lot of most geniuses, in some strange world of his own, full of nervous tensions of which we lesser mortals know

> nothing. The manifestations of these were such as to cause consternation to the weaker brethren.[38]

Here, John admits that Alan is a genius and begins with an insight and includes himself as part of "we lesser mortals" and then stresses the gulf with a more general "the weaker brethren," although John is literally Alan's "brethren" or "brother." Thus, John mixes psychological insight with a sardonic insight, which shows a tension between the genius and his own brother and the brotherhood of humans. John then turns to Alan's stutter and laugh and quotes from the Foreword of Mrs. Newman (Lyn Irvine) beginning "with ninety-nine people" and ending with "blue tweed suit" – a passage I discussed.[39] Another physical characteristic John comments on: "Alan's hands were those of a tormented man."[40] John also comments on his habit of deliberately losing his luggage and what John considers the farce of his "engagement" (John's quotation marks) to a female mathematician, which brings John back to "safe" women.[41] John is self-revealing when he calls Alan's fiancée "this unpromising female" and contrasts her with "the young, attractive and lively young women with whom I fell in love at intervals of about six months at a time" and John then recounts Alan corresponding with him for advice.[42] Sometimes in representing Alan's lack of empathy, John represents his own as can also be seen in the next small narrative within the larger narrative:

> If the episode of "the burglar" had not proved, ultimately, so fatal to Alan, I suppose this, too, might have been regarded as farcical. It occurred a couple of years before Alan's death and I believe that it was this that turned the scales against him. I had never had even the faintest notion that Alan was a homosexual.[43]

John muses on homosexuality and explains that Alan sent him a letter whose first sentence was "'I suppose you know I am a homosexual'" and John writes: "I knew no such thing."[44] John explains the story of the burglar who was someone at Alan's house who stole the gold watch Alan's father had left Alan and a few other items. John reports:

> Alan foolishly but typically reported the loss to the police, who did not seem much interested in 'the burglar.' But they were greatly interested in the prospect of prosecuting a don, a near-Olympic runner and a Fellow of the Royal Society for homosexual practices, then proscribed by the law.[45]

John raced up to Manchester and found a senior solicitor, who, unlike Alan's friends at the university there, persuaded Alan not to defend himself but to plead guilty so Alan would not face great publicity and would be put on probation and could keep his university appointment while undergoing medical treatment. Alan had to tell his mother in case journalists got hold of the story. John adds: "What exactly he told her or what she understood of it, I don't know: she did not seem to be greatly interested."[46] The brothers exchanged unpleasant letters. Alan accused John of not caring

> for his plight or that of homosexuals in general (the latter, perhaps, being not far wide of the mark) and that I was merely concerned to protect myself and my partners from adverse comment in the City from our Establishment friends.[47]

This accusation prompted John to reply: "It was so far from the truth that I sent him a tart reply of which I feel ashamed. It was a disagreeable business and the only occasion I can remember that we quarrelled."[48] If this were the only quarrel, then that in itself would be remarkable. John feels shame now. The brothers might differ on homosexuality, but they got along. Alan must have been under great pressure.

Two years later, during Whitsun, then a holiday in Britain, John returned with his daughter, after an evening at the cinema, and while out,

> the Manchester police had telephoned to say that Alan had been found dead in his house. Late as it was, I telephoned the ever kindly and shrewd Mr G, who promised to meet me at the station in Manchester the next morning.[49]

Mr. G was the solicitor who had handled Alan's burglar case. John praises the advice Mr. G. had given and the inquest itself, which had the verdict, which John, a solicitor, thought "considerate": "there were present some eight or nine reporters, some from the national press, with pencils poised and waiting for the homosexual revelations."[50] But none came. Alan's mother was in Italy and came home after the inquest, which John considered fortunate.

John does not hide the fact that Alan was upset with him two years before and reveals that his mother was also displeased with John after she returned:

> When my mother returned, she was highly indignant and made no secret of her belief that I had grossly mishandled the case. She evolved various theories of her own to establish to her own satisfaction that it was really an accident. But I had worked on the very same theories myself in Manchester for nearly three days and there was one fatal flaw in them. This was the half-eaten apple beside Alan's bed where his body was found.

> The apple was to disguise the bitter taste of the cyanide and thus ensure that the poison would do its work.[51]

The death of Alan Turing is a key moment in the narratives by Sarah and John, and their perspectives give a sense of this sudden event. For John, the time was sad and full of surprises: "In those unhappy days in Manchester, I visited Alan's psychiatrist who told me a great deal about Alan that I did not know before – among other things, that he loathed his mother."[52] This is a bombshell. The reader has no direct testimony from the psychiatrist, who also gave John two of Alan's patient exercise books and said there was a third book most likely at Alan's house:

> I viewed the two books in my hotel with horror, but I was still bent on proving the accident theory and decided I had better read them. I wish I had not. Alan had been a practising homosexual since the age of puberty. His comments on his mother were scarifying. To my great relief, I was mentioned only once or twice and not in opprobrious terms.[53]

These exercise books are not, as far as I can see, part of the Turing papers at King's College, Cambridge, and one notebook seems to have been left to Robin Gandy, but it does not seem to be accessible.[54] The third exercise book was found and returned to Alan's psychiatrist just two days before Sara Turing arrived at Alan's house looking "for clues bearing on her preconceived theories. I need hardly add that she remained unaware of the books and of Alan's feelings about her until the day of her death."[55] Not being able to see the notebooks, it is hard to know their contents, but taking John's word for it would suggest irony if not pathos. In his representation, the woman who idealized Alan, whom he had said, had "blind faith" in her younger son, and who fought so hard to secure his legacy, to make the man into a myth, a genius in the service of the nation, an inventor who help to make a world ready for computers, this woman, this mother, was loathed by the very son for whom she fought so hard. The human world is all too human.

It is hard to know John's feelings for his mother. He corrects and criticizes her but puts himself forward as someone who protects her and praises her. At the beginning, John said that he wanted to be honest and seek a balanced truth, a full portrait of Alan. From this account, although John does not talk much about his father (who is not that present in Sarah's biography), John seems close to him:

> I once asked my father what he hated most in the world. He replied, without a moment's hesitation, 'Humbug.' My mother has dispensed a

> liberal dose of it to the general public. Looking back on my father and brother as I remember them, I think I owe it to their memories to put the record straight.[56]

John wishes to correct his mother's humbug and honour the memories of his father and brother by putting the record straight. If Alan is complex, so too is John even if their complexity differs. The father, Julius Turing, does not seem to receive prolonged attention in either Sara's or John's account. This "dehumbuging" (my term) means, as John says, that he represents "some less appealing features of Alan's character and habits," and John admits something about Alan: "He was a complex man and much loved by many."[57] John expresses regret and remorse concerning Alan: "Had he been better understood when he was young – and if I, among others, had treated him with more consideration – he might be alive today."[58] Here, John takes responsibility and highlights how Alan deserved "more consideration" from others. The death haunts John:

> Why he took his own life remains, and is likely to remain, an unsolved mystery. I take the view, myself, that the immense pressures on him at Bletchley for so many years had taken their toll and that, but for these, he might have weathered these later storms.[59]

Alan's death is for John as it was for Sara a mystery, but, unlike Sara, John accepts the death as a suicide. Like Sara, John has his own theory, his centring on the pressures at Bletchley, which is a possibility. As with Sara, with John a certain ambivalence remains between Alan the genius and Alan the eccentric:

> If I have dwelled too much on Alan's idiosyncracies, it is because these were, to me, the essential Alan, but to many who saw a different side of him, he is remembered more for his modesty, his generosity and his kindness.

John is contradictory as he thinks that neglect of Alan as a child caused insecurities that led to his eccentricities, but then he says that this eccentric nature defined Alan for him. There is a certain tenderness, and John ends as he began his account of Alan – positively because, as Sara did, John quotes others and ends his account with a person Sara also quotes; that is, Mr. O' Hanlon's notice in Sherborne's magazine in June 1954, which praises Alan as practical despite "a negligé appearance," loveable and generous, despite his achievement in mathematics and science.[60]

In the past ten years, John Turing's son, Dermot, has written books about his uncle Alan and related subjects, so the family discussion continues, but, for the moment, owing to limitations of space, I cannot pursue more avenues, which was also true on the ongoing interpretation of Pierre and Marie Curie in the Curie family. Andrew Hodges, a mathematician of physics from Oxford, wrote a biography about Alan Turing in 1983, and I focus on Hodges' "Author's Note" as a way of moving from the accounts of contemporary family and friends to later biographers. Hodges begins that note with the limitations of sources:

> For a figure in world history, there is very little source material from which to reconstruct a picture of Alan Turing – few original documents, and little in the way of published commentary. Secrecy and embarrassment of various kinds are partly responsible, but there is a paucity of information even where taboo subjects are not involved.[61]

A mathematician, Hodges understands that role and this he shares with Turing:

> One must recognise, however, that Alan Turing did not conduct his life as that of a figure in world history: he tried as best he could to continue the life of an ordinary mathematician. And mathematicians (as compared with literary or political figures, entertainers or spies) do not usually expect to be heard of or written about, whatever their contribution. They do not really expect other people to know what mathematics is, and are generally happy to be left alone.[62]

Hodges notes that compared to other mathematicians, Alan Turing has biographical material that is substantial. After noting the obituaries by Max Newman, Robin Gandy, Philip Hall and others, Hodges observes:

> One person was not satisfied with this assessment, and sensed that some other kind of recognition was due. This was Mrs Turing, who in 1956 embarked upon writing a biography of her son – an extraordinary development by any standards, in which a seventy-five-year-old Guildford lady, not hitherto notable for literary or social confidence, and knowing almost nothing about science, was left to piece together some of the debris from the wreck of the modern world's dream. Her Victorian values still unshaken, she retained a strong belief in the idea that Alan's work had been and would be for the benefit of humanity.[63]

Hodges sees that Sara wished to see another kind of recognition for her son, and despite her lack of experience, her writing of the biography was extraordinary, the theme being Alan's work was for the benefit of humanity. Besides placing Sarah in her time and place, Hodges suggests what Sara's biography could be:

> Her slim volume appeared in 1959. There was perhaps a better book in Sara Turing, one that would have been a genuine *memoir* which placed the death alongside the other mysteries (as they were to her) of what her son had spent his life doing

and Hodges chronicles the ways Sara's book comes up short, including her "apparent emotional detachment" while making the interesting observation that Sara "was surely trying to prolong in her mind the lately increased rapport with her son, by projecting it back into a past life of which she knew little" while being "obsequious" to people of office or rank.[64] Hodges is judicious and includes praise and critique:

> Yet there was one point his mother grasped that better-informed people could be too sophisticated to see, namely that in 1945 he had set out to build a computer. She stuck to this at a time when everything surrounding the subject was still suffused in embarrassment.[65]

Sara's grasp and stick-to-it-ness Hodges notes, and he adds: "Tiptoeing among the minefields – and perhaps anyone but her would have found it impossible to continue – she did, ultimately, stand up for him as few others would."[66]

Hodges shows how adept Sara was to navigate in a hard environment and stood by Alan as few would, but he is ambivalent about her. Hodges has many interesting things to say about Alan Turing, for instance, how Bletchley Park came to be less secret and gay liberation occurred: "Taking these developments together, the 1970s therefore made it possible for someone to see *who Alan Turing was*, in a way that no one in his lifetime (but he himself) could have done."[67]

In the summer of 1968, Hodges says he was "reading about cybernetics and Turing machines as a mathematics student" but adds that he chose to study "mathematical physics, learning about relativity and quantum mechanics from Roger Penrose as a postgraduate student from 1972 onwards, and trying to make a contribution to the Penrose theory of twistors."[68] This interest in physics and computing as well as in a just recognition of homosexuality spurred Hodges' interest in Turing. Hodges discusses the sources for writing about Turing and doing so recognizes Sara Turing: "So I turn now to the question of original Turing papers, where my first acknowledgement of a debt must be to Mrs Turing."[69] Hodges goes into some detail about Sara's contribution through her book, collecting letters, depositing them at King's College and setting up the A. M. Turing

Trust, all of which helped future biographers, in this case Hodges. Moreover, Hodges sets out other debts, including to the subject of his book, and observes: "Alan Turing presumably thought that eventually a machine would be capable of writing a book such as this."[70] Hodges worries not about whether machines can think "but the place of such 'thought' in the body politic."[71] I shall return to Hodges' "Preface," published in 2014, after my discussion of Martin Davis "Foreword" to the centenary edition, which appeared in 2012, in accordance with my discussion according to the chronology of when a text appears.

Davis opens his "Foreword" by naming Alan Turing's mother: "Sara Turing, a woman in her seventies mourning the death of Alan, her younger son, a man that she failed to understand on so many levels, wrote this remarkable biographical essay."[72] Davis uses epideictic rhetoric, the blame of failure, the praise of her "remarkable" essay in discussing Sara here. He tries to be balanced in his response to the accounts of Alan by Sara and John, which make up this edition:

> But Alan Turing was a thoroughly unconventional man, whose method of dealing with life's situations was to think everything through from first principles, ignoring social expectations. And she was trying to fit him into a framework that reveals more about her and her social situation than it does about him. Alan's older brother John trying to fill in the gaps he saw in his mother's account, also ends up revealing a good deal about his own attitudes.[73]

Besides qualifying the contribution of Sara Turing and John Turing, Davis seeks briefly to examine questions readers might have reading these accounts in this edition. Davis tells of how Alan Turing contributed to decrypting the military communications of Germany. Moreover, Davis, among his very clear, informed and helpful analysis, says:

> It's important to emphasize that although the 'machines' in Turing's 1936 paper existed, and were only intended to exist, on paper, nevertheless they represented a paradigm shift in the way people thought about computation. Computation was not only 'number crunching,' but was also the execution of algorithms dealing with any kind of data.[74]

In addition to noting that in the biography Alan's mother does not mention his homosexuality, Davis sees John's view of Alan's homosexuality as a disorder as being in vogue when he wrote about his brother, a different view of sexuality than the one today.[75] Davis discusses the engagement: "Although Sara Turing doesn't mention it, Alan was engaged to a young woman, Joan Clarke, a co-worker at Bletchley Park, for several months.

Joan was a very talented mathematics student who had been recruited to work as a cryptanalyst."[76] This engagement is something John Turing examines, so Davis is responding to John's negative view of the unnamed Joan (whom Davis names) in contrast with the attractive young women whom John says he brought home on the weekends: "Even allowing for the prejudices of the time, this denigration of a capable intelligent woman was truly appalling."[77] Davis then shifts to Alan Turing at Princeton:

> It was decided that Alan should spend some time at Princeton so Church and he could explore their common interests, and Max Newman wrote Church to see what could be worked out. Turing did spend two years at Princeton followed by a year back in Cambridge before the war broke out.[78]

Davis contrasts the systems at Princeton and Cambridge, the one requiring the PhD, the other not. He adds that Turing's "dissertation was an important contribution to mathematical logic extending Gödel's work on undecidability."[79] The Automatic Computing Engine (ACE), for which Turing wrote a plan, Davis discusses, noting: "After the war had ended, Alan was eager to help build a working Universal computer."[80] The peacetime bureaucracy doomed this project, as there was no wartime Winston Churchill to move matters along, and so Alan left to Manchester at the invitation of Max Newman. There, Alan Turing "interacted with it not in pushing the kind of advanced software development he had outlined in his ACE report, but rather as a user to carry out computations related to the biological problems in which he had become interested."[81] Davis explains what led Alan Turing to raise the question – "Can Machines Think?" – "was that he saw in his ACE a first crude approximation to a human brain."[82] Alan Turing was interested in a machine holding a conversation indistinguishable from a person doing so, but, for Davis, he "was far too optimistic about the task of programming computers to achieve a command of natural language equivalent to that of every normal person."[83] Davis discusses the burglary and observes: "Alan foolishly went to the police when a few items had been stolen from his house."[84] He notes that the oestrogen injections caused Alan to grow breasts. The death of Alan Turing is what Davis concludes with before recommending further reading.[85] Davis observes: "Sara Turing would have it that it was her son's slovenly habits that led to his getting deadly cyanide on an apple he was eating. John Turing was convinced that it was suicide."[86] The position Davis takes is as follows:

> there is reason to believe that Alan did take his life, and that moreover he had staged his suicide in such a way that it would be clear to friends

> what he had done, while to his mother it would appear as a vindication of all her warnings about his slovenly habits.[87]

Davis imagines that Alan, who liked the film *Snow White and the Seven Dwarfs* and used to sing the song of the wicked witch about dipping the apple to let death seep, might have sung this as he prepared the poison and bit the apple.[88]

I conclude with Hodges' "Preface" to the edition of his book in 2014. Among his observations 31 years after the first edition is the view that "Alan Turing played a central part in world history."[89] Part of his later assessment goes beyond the original one in 1983:

> it was his individual freedom of mind, including his sexuality, that mattered – a question taken much more seriously in the post-1968, and even more in the post-1989, era. But beyond this, the global impact of pure science rises above all national boundaries, and the sheer timelessness of pure mathematics transcends the limitations of his twentieth-century span.[90]

Turing was of his time, beyond it and for all time. Like Davis, Hodges is a mathematician and gives a good assessment of Turing:

> The same timelessness lies behind the central element of Alan Turing's story: the universal machine of 1936, which became the general-purpose digital computer in 1945. The universal machine is the focal, revolutionary idea of Turing's life, but it did not stand alone; it flowed from his having given a new and precise formulation of the old concept of algorithm, or mechanical process. He could then say with confidence that all algorithms, all possible mechanical processes, could be implemented on a universal machine. His formulation became known immediately as 'the Turing machine' but now it is impossible not to see Turing machines as computer programs, or software.[91]

Hodges also notes that with the internet, Turing's universality saved the computer industry in the 1980s because they did not require reinvention. Moreover, Hodges notes, "In 2000 the leading mathematical logician Martin Davis, whose work since 1949 had greatly developed Turing's theory of computability, published a book which was in essence just what Turing could have written in 1948" and Hodges says that Davis, whose view of Turing I discussed, explained "the origin of the universal machine of 1936, showing how it became the stored-program computer of 1945, and making it clear that John von Neumann must have learnt from Turing's 1936 work in formulating his better-known plan."[92] Like Davis, Hodges makes valuable assessments as mathematicians, and he also shows what we owe to Turing:

> Online search engines, which work with such astonishing speed and power, are algorithms, and so equivalent to Turing machines. They are also descendants of the particular algorithms, using sophisticated logic, statistics and parallel processing, that Turing expertly pioneered for Enigma-breaking.[93]

For Hodges, the passage of time has allowed for a proper assessment of the contribution of Alan Turing: "Only after thirty years did the scale and depth of wartime cryptanalysis at Bletchley Park begin to leak out, allowing a serious assessment of Alan Turing's life to be attempted."[94] Only after three decades did the biography have enough information. Sara Turing seems to have had an inkling of this situation, including the government secrecy, and did her best to write a biography and gather an archive for future biographers. She got things started. Hodges adds another point: "This point coincided with the break-out of cryptology theory into an expanding computer science, with a reassessment of the Second World War in general, and with the impact of 1970s sexual liberation."[95] Hodges talks about many topics in this "Preface" to this edition of 2014, but he says he cannot supplement, after all the new material, scientific developments and the edition of 1983 in a short space but that this prefatory matter is also a later looking ahead to 1983 in a book that had looked back to the family of Alan Turing and his birth: "The curious cocktail of topics in this Preface is also offered as an aperitif for the story itself, inviting the reader to travel back over more than a century, and to enter the world of 1911."[96] This is an afterword in time of composition but a preface to a new edition. It looks back in time and is a new frame to an old text. Hodges makes no apology for the Victorian roots he drew on and closes with the image of Alan Turing being a pawn on the mathematical chessboard of Lewis Carroll and with the romance of Walt Whitman's historical account of what is to come. The last sentence in the "Preface" elaborates on Carroll and Whitman as roots: "These dreams from the nineteenth century still speak to the crimes and follies of the twenty-first."[97] This is a thoughtful observation, the past speaking to the future, dreams calling out the criminal and foolish future. At the end of the Preface, Hodges has left Sara Turing behind, but earlier he does include the account that Mrs. S. Clayton, Alan's housekeeper about his death, which she wrote to Sara on 8 June 1954 and which describes how the police came to take charge of Alan's house.[98] Actions such as that taking charge adds to the mystery of the death of Alan Turing.

In the next chapter, I turn to Denis Kelly, an assistant of Winston Churchill in the late 1940s and the 1950s. Kelly, born four years after Turing, writes about thirty years after working with Churchill. Like Marie Curie's life of Pierre Curie, Kelly mixes self-representation with representation, autobiography with biography, memory with present reflection. Time becomes a central matter.

Notes

1 Sara Turing, *Alan M. Turing Centenary Edition with a Foreword by* Martin Davis *and an Afterword by* John Turing (Cambridge: Cambridge University Press, 2012), 114. First edition published by W. Heffer & Sons, Ltd 1959; Foreword to the First Edition © L. Irvine 1959, 2012 Foreword to the Second Edition © M. Davis 2012; Afterword © J. Turing 2012).
2 Turing, *Alan M. Turing*, 117.
3 Turing, *Alan M. Turing*, 118.
4 Turing, *Alan M. Turing*, 119.
5 Turing, *Alan M. Turing*, 119.
6 Turing, *Alan M. Turing*, 121.
7 John Turing, "My Brother Alan," in Sara Turing, *Alan M. Turing Centenary Edition with a Foreword by* Martin Davis *and an Afterword by* John Turing (Cambridge: Cambridge University Press, 2012), 145. An excerpt, which appears in *The Atlantic* by John Turing on JUNE 22, 2012 as "My brother, the Genius," begins: "My mother gives a true picture of Alan's generosity." The "Editor's note" in *The Atlantic* includes "John wrote this account in the '60s and '70s, in part to correct what he saw as inaccuracies in his mother's account. In the selection below, John considers his brother's personal side." See https://www.theatlantic.com/technology/archive/2012/06/my-brother-the-genius/258886/.
8 Turing, "My Brother Alan," 145.
9 Turing, "My Brother Alan," 145.
10 Turing, "My Brother Alan," 145.
11 Turing, "My Brother Alan," 146.
12 Turing, "My Brother Alan," 147.
13 Turing, "My Brother Alan," 147.
14 Turing, "My Brother Alan," 148.
15 Turing, "My Brother Alan," 148.
16 Turing, "My Brother Alan," 150.
17 Turing, "My Brother Alan," 150.
18 Turing, "My Brother Alan," 150.
19 Turing, "My Brother Alan," 150.
20 Turing, "My Brother Alan," 153.
21 Turing, "My Brother Alan," 154.
22 Turing, "My Brother Alan," 154.
23 Turing, "My Brother Alan," 155.
24 Turing, "My Brother Alan," 156.
25 Turing, "My Brother Alan," 156.
26 Turing, "My Brother Alan," 156.
27 Turing, "My Brother Alan," 157.
28 Turing, "My Brother Alan," 158.

29 Turing, "My Brother Alan," 158.
30 Turing, "My Brother Alan," 158.
31 Turing, "My Brother Alan," 158.
32 Turing, "My Brother Alan," 159.
33 Turing, "My Brother Alan," 159.
34 Turing, "My Brother Alan," 159.
35 Turing, "My Brother Alan," 159.
36 Turing, "My Brother Alan," 160.
37 Turing, "My Brother Alan," 160.
38 Turing, "My Brother Alan," 160.
39 Turing, "My Brother Alan," 161.
40 Turing, "My Brother Alan," 161.
41 Turing, "My Brother Alan," 161.
42 Turing, "My Brother Alan," 162.
43 Turing, "My Brother Alan," 162.
44 Turing, "My Brother Alan," 163.
45 Turing, "My Brother Alan," 163.
46 Turing, "My Brother Alan," 163–164.
47 Turing, "My Brother Alan," 164.
48 Turing, "My Brother Alan," 164.
49 Turing, "My Brother Alan," 164.
50 Turing, "My Brother Alan," 164.
51 Turing, "My Brother Alan," 165.
52 Turing, "My Brother Alan," 165.
53 Turing, "My Brother Alan," 165.
54 Mike Yates, "Obituary: Robin Gandy," *Independent*, Friday 24 November 1995 00:02 GMT, https://www.independent.co.uk/news/people/obituary-robin-gandy-1583406.html.
55 Turing, "My Brother Alan," 165.
56 Turing, "My Brother Alan," 165.
57 Turing, "My Brother Alan," 166.
58 Turing, "My Brother Alan," 166.
59 Turing, "My Brother Alan," 166.
60 Turing, "My Brother Alan," 166.
61 Andrew Hodges, "Author's Note," in *Alan Turing: The Enigma the Book that Inspired the Film the Imitation Game* edited by A. Hodges (Princeton: Princeton University Press, 1983, rpt. 2014), 666. Hodges' Preface is 2014.
62 Hodges, "Author's Note," 666.
63 Hodges, "Author's Note," 667–668.
64 Hodges, "Author's Note," 668.
65 Hodges, "Author's Note," 669.
66 Hodges, "Author's Note," 669.
67 Hodges, "Author's Note," 672.
68 Hodges, "Author's Note," 672–673.
69 Hodges, "Author's Note," 674.
70 Hodges, "Author's Note," 678.
71 Hodges, "Author's Note," 679.
72 Martin Davis, "Foreword to the Centenary Edition," in Sara Turing, *Alan M. Turing Centenary Edition with a Foreword by* Martin Davis *and an Afterword by* John Turing (Cambridge: Cambridge University Press, 2012), vii.
73 Davis, "Foreword," vii.

74 Davis, "Foreword," xi.
75 Davis, "Foreword," xii–xiii.
76 Davis, "Foreword," xiii.
77 Davis, "Foreword," xiv.
78 Davis, "Foreword," xiv.
79 Davis, "Foreword," xv.
80 Davis, "Foreword," xv.
81 Davis, "Foreword," xv.
82 Davis, "Foreword," xvi.
83 Davis, "Foreword," xvi.
84 Davis, "Foreword," xvi.
85 Andrew Hodges, *Alan Turing: The Enigma*; David Leavitt, *The Man Who Knew too Much: Alan Turing and the Invention of the Computer* (New York: W.W. Norton, 2006); Martin Davis, *The Universal Computer: The Road from Leibniz to Turing* (New York: Norton, 2000).
86 Davis, "Foreword," xvii.
87 Davis, "Foreword," xvii.
88 Davis, "Foreword," xvii.
89 Andrew Hodges, "Preface," in Andrew Hodges, *Alan Turing: The Enigma the Book that Inspired the Film the Imitation Game*, edited by A. Hodges (Princeton: Princeton University Press, 1983, rpt. 2014), xv.
90 Hodges, "Preface," xv.
91 Hodges, "Preface," xvi.
92 Hodges, "Preface," xvii.
93 Hodges, "Preface," xviii.
94 Hodges, "Preface," xviii.
95 Hodges, "Preface," xviii.
96 Hodges, "Preface," xxxi.
97 Hodges, "Preface," xxxii.
98 Hodges, "Preface," xxviii–xxix.

12 Denis Kelly, *Churchill Memories* and His Work with Winston Churchill

On 21 October 1941, crypanalysts at Bletchley Park, including Alan Turing, wrote Prime Minister Winston Churchill, asking for more resources. The confidential letter begins by referring to Churchill's visit to Bletchley: "Some weeks ago you paid us the honour of a visit, and we believe that you regard our work as important."[1] Owing to a lack of "sufficient staff," the "work is being held up" or "in some cases is not being done at all."[2] Apparently, Churchill granted this request.[3] Some from the 1980s onward claim that Churchill said that no one made more of a contribution than Turing did to the Allied effort in the second World War, but Jonathan Schilling says that:

> no documentation has ever been found in which Churchill specifically praises Turing, and the claim does not supply a date or a context in which this statement was supposed to have been made. Certainly, Churchill made no such declaration in public during his lifetime since the existence of Ultra remained fully under wraps until nearly a decade after his death.[4]

Churchill met Turing in a group, and there is no documented evidence that they ever had a private conversation or that Churchill singled Turing out in the war effort. Churchill himself understood the importance of science as well as of history, as Bertrand Russell did, but as there is no written evidence of any relation between Churchill and Turing. I return to Churchill through another connection, that with Denis Kelly.

In *The Words of Winston Churchill*, I talked about the verbal art of Churchill as a Parliamentarian but also more briefly as a historian and writer. In that analysis, I discussed his assistants and those who wrote about Churchill's history and writing.[5] Denis Kelly was one of Churchill's literary assistants and maintains that Churchill put his stamp on his writing even when later he had a team working with him.[6] As I said in the first of my three books on Churchill's language, especially in Parliament, Denis

DOI: 10.4324/9781003651871-13

Kelly deserved and deserves more attention: he has insights into Churchill. In this and the next chapter, I will reiterate some of what I discussed about Kelly in the book but will amplify that into a longer discussion, whose length made two chapters a more accessible option for readers. A student at Marlborough, Kelly later read history at Balliol and then trained in law. One of Churchill's literary assistants and his first archivist, Kelly admired the power of Churchill's language and style and his work as an author, and Kelly learnt from Churchill about writing and authorship. Kelly learned how hard it is to write clear and simply prose.[7] Churchill edited the work Kelly presented him, thereby teaching him to write, but Churchill knew he himself had a gift for writing in school but did not claim to be a great writer.[8] According to Kelly, Churchill stressed that facts needed to be presented well.[9] When Kelly was an assistant, Churchill had shifted from single author to master artist with his team. Kelly describes the way they worked. First, they established the table of contents. Then, they collected key documents and composed a connecting narrative. Each chapter was about 5000–8000 words and was circulated to experts for comment and reworked at least six times. For Kelly, Churchill considered chronology to be the key to narrative, although there may be some subjects breaking in but with the clearest and fewest words.[10] Kelly says that Chartwell and the support of family were earned in long nights of writing, checking and rewriting of fifty books and many newspaper articles.[11] Moreover, Kelly likened Churchill as an author to a master chef, not cooking the entire banquet but, he implies, being responsible for it.[12] Kelly brings the team to life, for instance, in his portrait of "Mr. Wood, the proof-reader" who had proofread *The Life of Marlborough*, which Kelly praises as Churchill's best book, for Harraps.[13] Kelly says that Churchill did not sacrifice "historical truth to a good yarn"[14] and past leaders were to him as present leaders were.[15] Kelly, like Churchill's other accomplished assistants such as Maurice Ashley, provides good background on Churchill as a historian and writer, but Kelly also requires closer examination. Like Charles Eade and Martin Gilbert, I drew on Kelly in my study of Churchill.[16] Over the decades, I worked at the Churchill Archives Centre, and the Directors, principally Piers Brendon and Allen Packwood, have been exceptionally helpful.[17] In focusing my work from a general discussion of Churchill's language, I consulted the manuscripts and papers of William Deakin and Denis Kelly. Originally, my study of Churchill was to take a detailed look at him as a historian but not focus on that exclusively as Ashley does. Thanks to Allen Packwood, I came to discuss the speeches in Parliament but not concentrate on them as historical documents and political instruments as David Reynolds does.[18] It was Allen Packwood who introduced me to the family and friends of Denis Kelly, and for years I thought how I could tell more of Kelly's story as I thought his *Churchill Memories* worth publication. Short of that, I could say something about Kelly and this and his other work, and here I shall say something more.[19] Kelly's memories of Churchill are part biography and part autobiography: the memoir is both of Kelly and of Churchill, his presence and voice coming through Kelly's memory and the prose with which he shapes it.

Later in life, Kelly hoped to publish *Churchill Memories* and a manuscript, "'War & Witchcraft' – about India & Burma 1940–1945," he refers to in a letter to Nicholas Bagnall.[20] Kelly ends the letter, on the second page with an appeal to Bagnall: "I am recovering from a dose of radiation to cure cancer of the throat, but I am on my feet & would so appreciate any help or criticism."[21] The hopes and courage of Kelly are evident here. Ill and receiving treatment, he wishes to publish his work. This is a poignant, human situation.

Kelly's papers and work with Winston Churchill require further attention. Here, I shall focus on *Churchill Memories*, a work that should be published as a book. For now, my analysis may help to bring it more to light. My book is about human connections, intellectual affinities and other associations in science, the humanities and related fields like history, literature and politics from the nineteenth century onwards. Moreover, my book is often about representation and self-representation. Like Churchill, I have tried to stick, more or less, to the chronology of when the subjects under discussion were born. Kelly's work is important in and of itself, but his connection with Churchill may be of particular interest.

"Chance Encounter," Chapter 1 of *Churchill Memories*, begins: "One can seldom foresee, and still less seldom influence, the turning points of one's life."[22] Kelly tells of his return from the Second World War and his start in law, with the help of a friend and a colleague of his own father.[23] Kelly shifts to the craft of the writer, quoting P. G. Wodehouse about where to begin a story, warning of fooling about with atmosphere or getting off "like a scalded cat" and leaving the public "at a loss."[24] Thus, Kelly introduces himself:

> I must stop writing like a scalded cat and explain that I am an elderly barrister who by chance catalogued Sir Winston Churchill private papers and helped him write his six volumes on the history of the Second World War and his four volumes on a history the English Speaking Peoples – all this between 1947 and 1957.[25]

Kelly is looking back decades to his work with Churchill, but he shifts to his own birth and early life: "I'm told I was born in Dublin Castle three months before the Sinn Fein Rising in Easter 1916 and I probably owe my life to my Irish nanny."[26] He describes the dramatic scene and concludes: "Soon afterwards, my Irish father, who was an intelligence officer in the

British Army, was put on the I.R.A. Death List and ordered back to England."[27] Kelly is adept with narrative, showing how the threads of his life come together. He elaborates on his father:

> MR. Lloyd George had sent him to Ireland with a large sum of Secret Service money to persuade the Irish to stop fighting each other and join the British Army and fight the Germans instead. 'I refuse to go' said my father 'unless there is somebody with me to keep the accounts, or I'll be accused of spending all this cash on women and drink'; so he had been alloted a discreet young man from the Inland Revenue.[28]

"MR." Kelly adds in handwriting in black ink to the typescript, and he strikes out – and one cannot see what is struck out but he adds in the margin in that deep black ink "to stop fighting each other and." Kelly says that no one would suspect a tax official of being an agent and notes that "as the whole world knows, their mission failed."[29] Still, Kelly says, his father was so impressed by this official that he urged him to leave the civil service and be admitted to the Bar. This tax lawyer became eminent and defended Wodehouse in a tax case, and it was to Raymond Needham that Wodehouse dedicated the volume whose opening paragraphs Kelly quoted to begin his *Churchill Memories* and to whom he "had turned after nearly five years with mountain guns on the North-West frontier of India and fighting the Japanese in Burma."[30] Needham helped Kelly to the Temple, where he occupied a small, cold room. Graham Dixon, a barrister, came to see Kelly there and introduced himself:

> I am Winston Churchill's tax-adviser. He has handed over all his private papers to trustees and I've told them that they must have the papers catalogued. I have recommended you for the job – £100 a month and all expenses.[31]

This moment changes Kelly's life, and he has shown the connections that led to this change. Kelly says that was a lot of money and recounts that

> on the 14th of May 1947 I climbed out of the train at Westerham into an English; summer afternoon, all the more perfect after the darkness, the cold and the rationed food of the hardest winter since the end of the Second World War. The imminence of the Great Man seemed, in my heated imagination, to extend to the tiny railway-station which is now extinct.[32]

Churchill, "the Great Man," he imagines. Kelly describes the journey in silence with a quiet taxi driver from Westerham to Chartwell, passing

Wolfe's statute and Quebec House before turning into a country lane before arriving at Chartwell five minutes later. Kelly then represents the house, the brick walls and crimson rhododendrons. He gives his impression of the house, likening it to an army headquarters, "falsely" reminding him "of the French chateaux from which the Generals of the Great War had safely and comfortably conducted the slaughter of the Western Front."[33] What matters is Kelly's impressions and not simply an accurate depiction of Churchill and his surroundings: Kelly wants to make a false impression part of his self-representation and his representation of Churchill and his environment, thereby showing his own fallibility. The inside of the house Kelly describes in telling detail: "Oscar Némon's bust of Churchill stood on a table in the entrance-hall beside the visitors' book which I never signed."[34] The reason for not signing is mysterious. Did he not like to follow rules and be conventional, or did he not think he was important enough to sign? Kelly continues the description of the house: "To the left was the library, with Lord Randolph Churchill's manuscript letters piously preserved in blue, leather-bound volumes to the left of the door, and filled from floor to ceiling with all kinds of books except fiction."[35] The filial piety Kelly stresses as well as Churchill's love of books except for fiction, although Churchill did write a novel, *Savrola*.[36] In the description, Kelly shifts from past to later in the present of the writing to a framed map above what was once the fireplace, the map being "the floating 'Mulberry Harbour' which nourished the Anglo-American landings in Normandy in June 1944."[37] Here, past and present, private and public, story and history meet in memories. Kelly also quotes Churchill himself, but not at the original meeting that Kelly is leading to, but a conversation sometime later: "'Many years before' be later told me 'I proposed it in a Cabinet paper, but by a bit of luck it wasn't published and the German General Staff never got hold of it.'"[38] This quotation provides drama while delaying the description of the present, a kind of positive narrative tension. Kelly looks back while he looks ahead, being conducted down the corridor to the office of his secretaries and upstairs where he presented himself for tea.[39] He says this was a unique meal – a pudding from the United States at a time of rationing – in the house of Lady Churchill, although he ate many meals there in the ten following years.

Kelly continues with shifts in time. He comes next to a portrait of Churchill: "He was shorter than I expected, with a big head, an undistinguished nose, fierce blue eyes, broad shoulders, butcher's wrists and extraordinarily long, sensitive fingers – as I later realised when I watched him write and paint."[40] Here, looking back, Kelly begins this description with "he" and not "Churchill" and mixes his first impression with a later observation about Churchill's fingers. Kelly continues:

> We sat on the first floor in what was then the drawing-room and I admired a white chestnut tree in full blossom which towered below us.

> 'A noble vegetable' he remarked, sipping a very weak whisky and soda, and then led me to the basement, unlocked a couple of doors to a cellar beside the central-heating plant and said with a wave of his arm 'Your task, my boy, is to make Cosmos out of Chaos'.[41]

The tree in bloom, Churchill's words, and a description of Churchill's drink all blend in Kelly's prose. How did Kelly know at this point that the whisky and soda was weak? The wit of Churchill is ironic, a godlike task for a young assistant. Kelly then ends this depiction of his first meeting with Churchill:

> The dungeon was stacked from floor to ceiling with the black, tin deed-boxes used by old-fashioned Victorian solicitors and on the other side of the furnace was a smaller cellar crammed with copies of his books and still more boxes and box-files. Then he took me upstairs, ordered me a drink and a taxi and said goodbye.[42]

There is an abrupt and succinct description of the departure and farewell after a detailed and compressed account of the working space in the basement. There is, in this memory, a sense of Churchill and how the house was at this time, what was then the drawing room and how Churchill had the work materials organized for the archiving in a house where more research would be done and more books would be written.

Kelly says that he returned from London the next morning, collected the keys from the office of the secretaries and then went down into the cellar, where alone he contemplated "the chaos" before him.[43] Here, Kelly changes "task" to "chaos" to echo Churchill's word. Kelly has a sense of timing and narrative:

> 'Time spent in reconnaisance' they used to say in the Army 'is seldom wasted' and I felt the need to absorb the atmosphere, to touch and smell the despatch-boxes, the loose files, the black tin deed boxes, the bundles of documents, the heaps of press-cuttings and the vast kitchen table which filled the centre of the room. It was a pleasant place.[44]

Here is a long, vivid, detailed descriptive sentence then a brief sentence to stress how pleasant the space. He describes the lawn and sunlight outside through the barred windows shining on wooden racks that remind Kelly of a cupboard in his home in Sussex. He gives a sense of the archive:

> As I peered and peeped into various receptacles I realised I was handling the personal papers of a man who had experienced and shaped the history of the world in the last fifty years – school exercise-books from the age of nine; letters and despatches from the wars in India, the Sudan,

> Cuba and South Africa; file upon file about the Dardanelles, the Somme and Czarist Russia; secret telegrams from Roosevelt and Stalin – all entangled with the politics between the two World Wars and private family letters and bank-accounts. Everything had been kept and nothing had been thrown away and all were unique and perishable.[45]

The realization is that Kelly is handling the papers of Churchill, someone who experienced and shaped history over the previous half-century. The challenge was that each document was unique, and there were many papers. Soon Kelly realized that he needed help, and Cicely Gemmel and he created a system to catalogue the collection. Kelly dictated and she typed, and they could not read the documents as the lawyers needed them. Thus, all that was possible to record was the subject and first and last date of each bundle:

> Churchill had told me to concentrate on the year 1934 onwards i.e. the year when Hitler established his power and position in Germany, but even by working five days a week it was four months before our task was completed.[46]

The pressure of time, Kelly explains, came from the lawyers needing to prove the Trust genuine and from Churchill who needed key papers to write his "memoirs" of the war.[47] Despite this challenge, almost in archival terms like an emergency or rescue dig in archaeology, Kelly has some pleasant memories:

> it was a beautiful summer and after lunch with his other secretaries in the village we would doze for half an hour beside the goldfish pool to the trickle of water – falls and the squawks of the swans on the lake.[48]

The manuscript has directions for the printer, and here Kelly asks for a gap. He then sets out his method, relating his own experience in the war with that of his work for Churchill:

> The less you know, the less you can give away. In Burma we were told as little as possible about the plan of campaign or the composition of the Fourteenth Army, and we made it a rule in my battery to know no more than the identity of the infantry battalions we supported with our mountain-guns and the names of the places we had to attack. So there was little we could tell if captured and tortured. I followed the same rule when dealing with Churchill. I only read the documents needed for his war memoirs. Other papers, such as private family letters, bank

> accounts, and correspondence about the Duke of Windsor were none of my business, and having noted their date and subject-matter were returned unread to his archives. Any notes I made of his conversation were confined to topics he might wish to publish in his books and once used were either destroyed or locked into the fire-resistant cabinets. Keeping a diary would have been as disloyal as writing-up a love affair.[49]

Kelly shows discretion. Only after Churchill's death does he seem to have wished to publish these memories in which he lists some of the subjects he did not read about. Kelly implies that his memories are not the same as a diary at the time. He explains:

> My memory of anecdotes and incidents may thus be faulty but these precautions diminished the risk of indiscreet gossip outside our small, intimate circle, and if journalists tried to pump me there was little I could or would tell them.[50]

Kelly reminds the reader about the risks of leaked secrets and the need for trust and clearance, even if informal:

> Perhaps this was why I was never – to my knowledge – interviewed by the British Intelligence Service. I was, after all, let loose on the original texts of the most secret telegrams between Churchill, Roosevelt and Stalin, and Churchill often revealed his hopes and plans and speculations in unguarded conversation at the dinner-table. What if I were a Soviet spy? A few facsimiles, a few telephone calls and the new enemy would be kept in touch with the day-to-day workings of his mind and made privy to as yet undiscovered wartime secrets, but I have no doubt that the British secret service knew my pedigree.[51]

Kelly is adept in showing his access to the most secret of documents that would be of interest in the Cold War and says that the secret service would know him, perhaps because his father was an intelligence officer. Kelly takes up this point:

> My father had worked in the famous Admiralty Room 40 throughout the Great War; Admiral 'Blinker' Hall – so called because of his rapid-blinking eyes – had been the Director of Naval Intelligence and was my godfather; Marlborough and Balliol had failed to contaminate me with the lunatic ideas and ideals of Communism and Fascism in the 1930's; and every year till he died in, I think, 1944, the Admiral sent me a ten-shilling note as a birthday tip.[52]

Kelly's father was well-connected and had links to the intelligence community, including with Admiral Hall. The work did have its challenges in regard to secrecy:

> But at the end of my ten years with Churchill I was left with a problem. In the course of composing, correcting and revising his history of the Second World War and circulating copies for comment by his experts, many copies of the secret telegrams and minutes had been printed in galley-form and were now bursting the cellars.[53]

To solve the problem, Kelly suggests burning these duplicate documents, but Mr. Vincent, the head gardener, comes up with a better idea – burying them. Kelly once moves back and forth in time: "I am glad to record that Mr. Vincent and I shared the small section of St. Paul's Cathedral reserved at Churchill's funeral for those, like us, who had worked most closely at his home."[54] Kelly, at the time of the writing, remembers the time at Chartwell and then the funeral.

Kelly was worried the house would burn with all the papers, and the solicitor to the Trust, Anthony Moir, arranged with Chubbs to install some "fire-resistant cabinets."[55] Kelly was anxious to copy the key papers, but fortunately they found a folder "Master File."[56] Moreover, Kelly revisits Churchill's gift for writing, his respect for facts and the pains he took, as Churchill said, "'in dishing them up.'"[57] Kelly elaborates on the following sentence: "The first time I drafted a page of his war memoirs was a humbling and instructive experience."[58] Churchill asked Kelly to condense an expert account of the Blitz, which Kelly reduced from 150 to 3 pages, which Churchill corrected with his red pen as if he were, in Kelly's image, a topiarist: "In the middle of this penitential process he gently turned to me and said: 'I hope you don't mind my doing this?' 'Sir' I answered 'I'm getting a free lesson in writing English.'"[59] Kelly uses dialogue to make the sense more dramatic and literally gives voice to Churchill and him while showing the gentle mentor at work. The experience is something Kelly prizes:

> Then came a great privilege. When cataloguing and indexing two cellars' of papers we had been able to send key-documents upstairs where he toiled on his War Memoirs with a brilliant, charming and anything but donnish don named Frederick William Deakin – D.S.O., First Class Honours in Modern History at Oxford – who had parachuted into Yugoslavia as one of Churchill's personal emissaries to Tito.[60]

This is no ordinary workplace: Deakin is proof that Churchill surrounds himself with extraordinary people. Kelly praises Churchill and Deakin, whose experience in Yugoslavia he represents dramatically. Deakin, as Kelly reports, was offered to head St. Anthony's College, Oxford. He makes the consequence dramatic by reporting a dialogue between Churchill and him, after Churchill threw him a cigar, and Churchill mentioned that Deakin would not be able to come down from Oxford so much, and he concludes by asking Kelly: "Would you like to stay on and help?"[61] Kelly indents this dialogue for emphasis, and he then makes a paragraph of the following sentence: "Thus started the most exacting and exciting ten years of my life."[62] Here, Kelly shows his awareness of style, balancing "exacting" with "exciting." After discussing unfairness and quoting "The Prof" – that is Lord Cherwell – on injustice, Kelly lauds Churchill for understanding how unjust life is: "Churchill knew this very well. In all the long years I was with him I never once heard him complain that his dismissal by the British electors at the end of the war was unfair."[63] Although Kelly also praises the humility of Churchill for seeking out experts or specialists, he notes that Churchill "liked to get his own way."[64] But Kelly states that Churchill heeded criticism and listened to comments and made corrections of fact – "he never sacrificed historical truth to a good yarn," and he has struck out "mythical" for "good."[65] Kelly considers Churchill's many gifts:

> Courage, respect for the truth, however unpalateable, attention to detail, the ability to work hard and to concentrate for lengthy periods of time – all these were some of the facets of Churchill's genius, but what most impressed me was his gift of going to the heart of the matter, of discerning what was vital in a complicated and seemingly overwhelming event.[66]

The most impressive part of Churchill's genius was his ability to go to the nub, and he gives several examples from the Second World War. Kelly elaborates:

> I once asked him: 'What were the biggest mistakes of the war?' He replied at once: 'Losing Singapore and letting the Russians into Europe'. The collapse of the British, French and Dutch Empires led to millions of innocent deaths in the Far East and is still doing so.[67]

Kelly continues his analysis with speculation and observes:

> "Stalin", he was fond of repeating "is reported to have said: 'The death of one man is a tragedy. The death of a million men is merely a matter of statistics'." He was right, of course, in his judgment of what was happening in the Soviet Union after the war and what was to go on happening till Josef Stalin died, and we are now told that Stalin killed more Russians than Hitler, but all this was a dinner-table dream born from the frustration of helplessly watching the Iron Curtain and the iron grip

> of the most blood-thirsty of modern Czars destroying so much of what he had hoped and fought for.[68]

Kelly brings out the tension between Stalin and Churchill, whom he quotes. This is also a seismic change in Europe and the world after the war. Once more, Kelly is able to bring together the private and the public. His interpretation involves Churchill fighting tyranny and evil.

Kelly discusses Churchill as a great author, which takes hard work to make something clear and readable, in the context of the way Graham Greene, P. G. Wodehouse and Tolstoy revised painstakingly. For Kelly, non-fiction has demands beyond fiction:

> The task is harder when describing Facts instead of Fiction because an imaginary tale, once started, can take on a life of its own and the story writes itself, whereas Facts obstinately refuse to be clear and simple, perhaps because the truth, as someone remarked, is seldom pure and never simple – which is why the world still needs so many politicians and policemen and lawyers to discover it.[69]

Kelly, trained in history and law, understands how hard it is to ascertain the facts and discover the truth. He asks a complex question:

> How do you portray the changing shifts of a battle, the to-and-fro of a Parliamentary Debate, or the perplexities and uncertainties and arguments which surround such momentous decisions as whether Alcibiades was right to invade Sicily, or Napoleon was right to invade Russia in 1812, or Churchill was right to ally Great Britain with the Bolsheviks in 1941 when we could have stood on the side-lines and left Stalin and Hitler to destroy each other?[70]

These points of query in this question are all good and hard to answer. Having written history and three volumes on the words of Churchill, especially in Parliament, I can say that answering difficult questions is sometimes harder after than when starting a book or project. And it is with this question that Kelly ends Chapter 1. He provides a perspective different from others, including the other assistants of Churchill.

Chapter 2, "Home Life at Chartwell," continues to show Kelly's power of description. Here he represents the mixture of work and the domestic sphere:

> The day began at seven o'clock at night. I had to go straight up to his bedroom because he could hear the taxi arrive and was impatient for

> the latest proofs from London. A grunted 'Good evening, my dear'; a long silence while he pored over the white rectangular pages and then, five or fifty minutes later: 'See you at dinner'. This was a scramble. He had a manservant who filled his bath, laid out his siren suit, and helped him undress and dress.[71]

This is an observation of life befitting of a novel but actually part of life writing, a mixture of biography and autobiography, a little like Boswell writing about himself in relation to Dr. Johnson, whose life he was writing. This is all done from memory. The details are impressive even if memory plays tricks with eyewitnesses, let alone those writing years after. Kelly continues:

> I had less than twenty minutes to unpack, wash, shave, get into a dinner jacket and be at his elbow for a gulp of tomato-juice or sherry. We did this standing up. No slouching in arm-chairs for cocktails and cigarettes. We ate at opposite ends of the table each with our copy of the current text."[72]

Here, Kelly shows the work routine, efficiency and work even when drinking: the text was the thing. Drink was part of Churchill's life and life with Churchill:

> Sherry with the soup; champagne with the main course; port with the cheese; brandy with the coffee; each with our pen and scribbling pad, then back to his study where another table flanked with small china figures of Nelson and Napoleon gave a twenty-mile view in day-time of the Weald of Kent.[73]

Work is the focus despite the variety of drinks, and Kelly mixes day and night, imagining the view he had seen during the day as they were working at night. The detail gives a glimpse into how Churchill worked at this point in his life:

> A woman secretary with the evening's letters. 'Yes'. 'No'. 'Regret'. 'Show again tomorrow'. Then down to the real work of the evening. Silent typewriter behind us to our right; long, hesitant, brilliant dictation; corrections with red pen; re-type. 'Envelope for the printer. Let me see it'. Initials at the bottom left-hand corner; sealed with a damp sponge from an earthenware bowl. ('You can get cancer by licking stamps'). 'Make sure he gets it by the "Davey Jones" – his nickname for the Godfrey Davis car-hire firm – tomorrow morning' and so to bed.[74]

Kelly mixes reported speech with an alternation of details, of sentences of varying length. The rhythm gives a sense of Churchill's decisiveness and efficiency. This routine was efficient and pleasant. There were moments of leisure amid the hard work.

I can only give a sense of Kelly's *Churchill Memories*. He goes back to the first time this routine happened and provides some more telling details. Kelly pulls off Churchill's suit, and Churchill speaks, false teeth out, for twenty minutes about the work of the day. The next day, breakfast was in one's room, and Churchill read sixteen newspapers, and he expected Kelly to do the same and comment. The budgerigar perched on Churchill's head, the ginger cat on his lap, the poodle, at the foot of the bed, sleeping across Churchill's ankles.

The process of producing a volume was exacting. For instance, Kelly notes:

> The first task was to make a contents-table to give the new volume a structure and proportion; the next was to collect the key documents for each chapter; the third was to compose the linking narrative. Each chapter was drafted, printed, re-drafted and reprinted at least half a dozen times and circulated to a group of experts for criticism and comment.[75]

This is only part of an intricate procedure. Churchill prized compression, accuracy and clarity.

Kelly is also good at giving a sense of place. The following passage, among others, is one of the reasons why I think Kelly's account should be published:

> Lunch was of similar dimensions to dinner and was followed by a walk round the estate. First the goldfish, fed by hand with live magots sent by post in fibre-packed tins; then down by waterfalls to the lake with the black and red New Zealand swans; up to the farm to inspect the pigs and then to the German prisoners of war felling a dead oak with enormous pride and enthusiasm as he watched from a camp stool. There were about half a dozen of them, strong, vigorous young men, unescorted so far as I could see by guards, armed or otherwise, and had waited for the Great Man to arrive before striking the final blows with axe and saw which brought it crashing into the bank. Then they fell on the trunk: and started lopping the branches with tremendous speed and accuracy. He sat viewing this remarkable scene in silence while I sat beside him on a log.[76]

This is a "remarkable scene." Nothing is ordinary even in the ordinary with Churchill. His love of animals and nature is apparent, and suddenly Kelly tells us of unguarded German prisoners of war who have a connection with Churchill on his estate. There seems to be a bond, trust between the great man and the prisoners in some kind of ritual with timing, ceremony and spectacle. From Blenheim to Chartwell, we have the remarkable Churchills. Born in the former, Churchill lives and writes in the latter, which, as Kelly reminded us before, Winston Churchill earned through writing. Churchill's writing supported his life in politics, especially early on when there was no proper salary for Members of Parliament. Churchill admired the Duke of Marlborough and his own parents, Randolph and Jennie. As great as Churchill was, he showed filial duty and love and revered his ancestors. Churchill was a meticulous writer who earned his way despite his aristocratic origins. Kelly gives a portrait of Churchill worthy of him.

Then Kelly provides a dialogue between Churchill and him. While Churchill had his afternoon sleep, Kelly and others toiled to give him enough work for the evening. Kelly is adept in mixing the private and larger public scene:

> Between 1947 and 1959, when the Russians laid siege to Berlin and Korea was invaded and the Americans exploded their nuclear weapons in the Pacific and Hungry fought for her life, his resolution and courage never faltered and the only time I saw him really depressed was a hot summer when the Chart well which nourished the waterfalls and goldfish-pools ran dry. He seemed to feel that the spring of his own home-life was drying up too.[77]

Kelly represents world affairs but also connects Churchill to the water on his land. As nature was drying up, so was the spring of his home life. Kelly did learn from Churchill and is a writer in his own right. Part of Kelly's account amplifies his description of water and his walks with Churchill, whose words he quotes, including on the music of the waters, as Churchill fed the fish. For instance, one quotation from Churchill on a walk is "Strange to think that I've earned all this with my pen" – then Kelly adds: "In a letter to Lawrence of Arabia congratulating him on "The Seven Pillars of Wisdom" he had dismissed his own account of the Great War in the six volumes of "The World. Crisis" as "dictated journalism" – then Kelly tells about the legend Handel writing "The Messiah" to pay his debts and says that Lord Birkenhead told Kelly's father "that the best work is done under pressure."[78] Kelly describes Churchill feeding the fish, birds and pigs. Moreover, Kelly tells stories about Churchill's pets, Toby the budgerigar, the proof mutilator who in Monte Carlo "flew out of a window of the Hôtel de Paris and was seen no more" and "Rufus I, the poodle who was run over on the on the sea-front at a Conservative party Conference at on the sea-front at a Conservative party Conference at Brighton" and

"Rufus II, another poodle who appears in a signed photograph I still possess of the three of us, plus our Naval expert, seated at the luncheon-table at Chartwell."[79] Kelly also describes how Churchill provided a searchlight and devised explosives to deter the fox who killed his swans. The water came back to the estate, and Kelly reports that it flows still.[80]

Chapter 3, "Literary Devils and the History of the Second World War," begins dramatically: "Churchill had very definite opinions about food and about growing old" – changing "He" to "Churchill" in a revision to the manuscript.[81] At the beginning of the chapter, I framed *Churchill Memories* with observations from throughout the manuscript, so I will avoid repetition, here concerning Mr. Wood, the fine proofreader. Churchill called on the experience of Sir Henry Pownall and Commodore G.R.G. Allen and put his own stamp on the prose making technical matters accessible.

Kelly turns to his own "fortunate" education, his love of history, his two schoolmasters at Marlborough and his history tutor at Balliol College, Oxford, his own study of the Middle Ages for three years, among other things and his military service in artillery in India and Burma during the Second World War. This helps to establish Kelly's credentials in his discussion of Churchill and history. Here, Kelly mentions Churchill being self-taught in history and how real leaders past and present seemed to him, something I discussed at the beginning of the chapter.[82] Past and present were important to Churchill when making decisions:

> When deliberating with the British and American chiefs of the armed forces who were to cross the Inglish Channel and liberate Trance in 1944, he baffled them all by asking which month Willian the Conqueror chose to launch his invasion of Britain.[83]

Kelly discusses how the problems of war stay the same even though the weapons change, and one of the illustrations he gives comes from Churchill in one of their conversations:

> Do you realise' he once asked me 'why the casualties in the Second World War were so much lighter then the casualties in the Great War? Because the weapons were more deadly.' The more I pondered this astonishing remark the more I realised its truth.[84]

Kelly connects Churchill's "you realise" with "I realised" to show the bond of realization or recognition (anagnorisis) between them. Kelly explains his work routine:

> In day-time I usually worked in the library on the ground floor. Here, in this friendly, high ceilinged, quiet room I could finally assemble the

> documents, the preliminary draft and the comments and criticisms of the experts on whom we relied for the accuracy of the impending chapter until summoned to the study upstairs.[85]

Kelly describes Churchill's study in detail, a description of interest, part of which I quote here:

> It had the same atmosphere as the Cabinet Room at 10 Downing Street. An uncurtained archway led to his bedroom – he worked in bed in the morning and at his writing-table at night – and, like No, 10, both rooms had an air of sunlight, leather-bound books, antique furniture and immediate and insistent contact with the real world outside.[86]

The space, atmosphere and routine Kelly compares to 10 Downing, the private house and the public residence. Kelly's account of the work routine is also instructive and shows the human side of Churchill, for instance:

> Brain work makes you just as hungry as manual work. After labouring with pen and papers till two o'clock in the morning I would fetch him the bowl of clear soup kept in the pantry across the landing from his study. He would scoop it up; I would see him into bed and listen in darkness to his reflections on the work we had done that evening and the work which lay ahead next day.[87]

The hard work in a domestic space, the mixture of the oral and the written, the bond of conversation combined with dialogue, which Kelly also records, gives a sense of certain moments of Churchill's life and life with Churchill. Churchill gives Kelly whisky and a cigar to cure him of a flu, which Kelly describes at some length and with dialogue, all the while work continued:

> I had distilled the secret reports sent to him during the war on how, months before they were launched, the British scientific intelligence services had assembled an accurate picture of the German flying-bombs and rockets and had proposed the measures to counter them, all written by a scientist named Professor R.V. Jones whose mastery of English was equalled by his mastery of the subject. The tale was as dramatic as a detective-story and quenched all thought of illness or fatigue and in due course bore fruit in three of the most fascinating chapters in his account of the Second World War.[88]

Kelly crossed out "enthralling" and added "dramatic." Kelly represents and stresses the dramatic throughout his account. One of the contributions that Kelly makes in *Churchill Memories* is his observations of the life of

Churchill at this time as it centres on his writing of books, and in this chapter, Kelly gives an account of the making of the volumes about the Second World War. He gives credit where it is due as in the case of Jones (1911–1997), a distinguished physicist, five years older than Kelly (1916–1990).

Kelly encouraged Churchill to give a more detailed account of the Yalta Conference in February 1945 even though it was a painful memory for Churchill. Kelly's account provides important details, including this passage: "In a brief, typewritten note dictated some months before I tackled the British record, Churchill had summarised the Crimea Conference in a single sentence: 'How I fought for Poland'. He expected and received no thanks."[89] As can be seen in the last two sentences, Churchill's plain style here seems to rub off on Kelly. Interpretation is another contribution that Kelly makes:

> His concern for the Common Man dominated & diminished the sixth and final volume of his war memoirs, which described their tremendous events. The year was now 1951; tension between the Soviet Union and her allies and the rest of the world was tightening day by day; he kept repeating 'I don't want to do anything which will make things worse.' And so he became more cautious about what he wrote. But some things had to be said. He had failed to keep the Russians out of Europe; he had failed to keep the confidence of the British electorate and had been driven from power in the very middle of the last, decisive summit conference with the Soviet leaders at Potsdam; having fought the war, he had been denied any part in making the peace. And so we had the Iron Curtain.[90]

For stress, Kelly adds in black handwriting "& diminished" as well as "which described their tremendous events" to show his judgement that this "concern" was not felicitous while also adding "now" for emphasis before "1951." Kelly emphasizes Churchill's failures and their consequences. He tries to be discreet with Churchill: "He never talked to me about losing the General Election and I thought it wiser not to ask him why he had ever called it."[91] Kelly does mention that historians consider that to be a blunder in 1945, but he also notes that Churchill valued Clement Atlee's loyalty during the war and Kelly blames himself for not getting the draft done to stimulate Churchill to write a tribute to Atlee.[92] According to Kelly, Stalin knew where he stood with capitalists like Churchill and Lord Beaverbrook but less so with "deviationists" like Atlee and the Socialists he led.[93] Kelly considers the question of Churchill and the Soviet Union: "As for keeping the Russians out of Europe, he largely contented himself by publishing his telegrams to Roosevelt and Eisenhower begging them to push their armies as far East as possible before the war ended."[94] Kelly sets out how strong the allies were in relation to the Soviet Union, including having an atomic bomb. "'I would have said to the Russians' he told me 'Go back to your own country or there'll be something worse coming to you.'"[95] Kelly punctuates his account with the words of Winston Churchill and interprets the situation:

> One will never know, of course, if he could have persuaded the United States leaders to support such a threat, but knowing him as I did, I am sure he would have had a jolly good try. The tragedy of the Second World War was that when the United States woke up to the Soviet danger, it was too late to do much about it – except re-arm, and to have said all this too bluntly in the final volume would have done more harm than good. He was Prime Minister for the second time when it was published and for the sake of world peace he kept in on a low key.[96]

For Kelly, Churchill would have tried hard to convince the United States, which woke up too late, and he moderated his writing when he was Prime Minister for the second time. Kelly records Churchill's views of the threat of a Third World War and calls attention to the limits of his own remembrance:

> 'I write all this from memory – which is fallible and his ideas which I have here set down are now, one hopes, commonplace, but when brooding over these grim realities he always returned to a second theme – 'This nuclear business must not make us defeatist'.[97]

Kelly continues with a sympathetic analysis of Churchill, who favoured active resistance to tyranny:

> He was deeply hurt at being called a war-monger; he had, as I have already said, analysed and described in eloquent detail the blood-bath of Paschendale and the Somme; he had fought in the front line trenches; he knew at first-hand what war was like; he would do all he could to prevent another catastrophe; but if he failed, he was quite prepared to die himself. Searching for peace was one thing. Giving way to a bully was something quite different. If the peoples of the world had stood up to Hitler, there would have been no Second World War. If they didn't stand up to the Soviets there could be a third one. All very obvious, but people dislike being told the obvious if the obvious is unpleasant.[98]

Churchill had experienced war and worked for peace, but peace did not come from appeasement. He told people unpleasant truths based in how the world works.

For Kelly, a study of how historians write would be worthwhile. Kelly speculates on how Edward Gibbon, Thomas Carlyle and Thomas Babington Macaulay wrote, and says:

> Churchill, who never considered himself more than an amateur historian, used a variation of his three great masters' methods and stood, strode or sat as the mood suited him with an extremely weak glass of whisky and soda on his desk which was seldom finished at the end of a three-hour evening.[99]

Kelly humanizes Churchill in the context of the legends and possible methods of these historians whom Churchill admired. Kelly also gives a full and adept description of the process of producing the books:

> 'I've more or less lost the art of calligraphy' he explained 'and so I dictate'. A secretary sat at a card-table a few feet behind his desk and to his quiet right, and took down his words on a quiet typewriter. The sentences emerged very slowly, so slowly that they could have been written in longhand, with pauses and hesitations while he silently discovered the right noun, adjective or phrase which, when it came out, was so exact that you said to yourself 'If only I'd thought of it myself!'[100]

This phase is only part of a process that also involves correcting proofs and Kelly taking a chapter by train to London where a car took it to a printer, who then sent it to Churchill's home at Hyde Park Gate to be taken back by Kelly or, if he was unavailable, by train and taxi: so what Churchill finished at two in the morning would be ready for him at seven that evening. There were eight proofs for Churchill, Kelly, a spare, and others for technical experts and for Mr. C. C. Wood. Meanwhile Churchill pressed ahead with subsequent chapters until the first draft of the volume, which Kelly thought "a good rule" and explains why at some length.[101] Kelly leaves no doubt that the language and the literary style are Churchill's:

> Just as a Master Chef mysteriously puts his own stamp on the various dishes of his kitchen-hands and so produces the banquet, so Churchill mentally digested the contributions of his advisers and put his literary stamp on the final text.[102]

Kelly's image of the master chef, which I mentioned early in the chapter when framing *Churchill Memories*, ends Chapter 3, an assurance that the work is ultimately that of Churchill. In the next chapter, I shall discuss the second part of *Churchill Memories* in some detail.

Notes

1 The letter can be found at the National Archives as part of "A dramatic plea for Churchill's help from Bletchley Park," at https://www.nationalarchives.gov.uk/explore-the-collection/stories/a-dramatic-plea-for-churchills-help/
2 "A dramatic plea for Churchill's help from Bletchley Park."
3 "A dramatic plea," at https://www.nationalarchives.gov.uk/explore-the-collection/stories/a-dramatic-plea-for-churchills-help/
4 Jonathan Schilling, "Myths: Fact vs Fiction: Churchill: Turing Made the Single Biggest Contribution to Allied Victory," International Churchill Society, https://winstonchurchill.org/resources/myths/churchill-turing-made-the-single-biggest-contribution-to-allied-victory/[n p].
5 Jonathan Locke Hart, *The Words of Winston Churchill* (New York: Routledge, 2023), 24–26. My great thanks to Victoria Kelly for her kindness and permission to discuss the work of her father, Denis Kelly, and to Allen Packwood for putting us in touch; for his great encouragement, knowledge and wisdom, for which I am always grateful. I got to writing so much about Denis Kelly that I had to break my work into two chapters.
6 Denis Kelly, *Churchill Memories* [unpublished manuscript], Churchill Archives Centre, Deke 2.
7 Denis Kelly, "Home Life at Chartwell," *Churchill Memories* [unpublished manuscript], Churchill Archives, Deke 2, 11.
8 Kelly, "Home," 12.
9 Kelly, "Home," 12.
10 Kelly, "Home," 4; see 3.
11 Kelly, "Home," 7.
12 Denis Kelly, "Chapter 3. Literary Devils and the History of the Second World War," *Churchill Memories* [unpublished manuscript], Churchill Archives, Deke 2, 3.
13 Kelly, "Literary Devils," 3.
14 Kelly, "Literary Devils," 7.
15 Kelly, "Literary Devils," 14.
16 Maurice Ashley, *Churchill as Historian: Churchill by his Contemporaries*, edited by Charles Eade (London: Hutchinson and Co. Ltd., 1953); Martin Gilbert, *Churchill: A Life* (London: Heinemann, 1991). Jock Colville, Churchill's private secretary, was also important to him. See Warren Dockter, "Managing a Giant: Jock Colville and Winston Churchill," in *Private Secretaries to the Prime Minister Foreign Affairs from Churchill to Thatcher*, edited by Andrew Holt and Warren Dockter (London and New York: Routledge, 2017), 8–33.
17 See, for instance, Piers Brendon, *Winston Churchill: A Brief Life* (London: Secker & Warburg, 1984) and Allen Packwood, *How Churchill Waged War: The Most Challenging Decisions of the Second World War* (Barnsley: Frontline Books/Pen & Sword Books, 2018).
18 See David Reynolds, *In Command of History: Churchill Fighting and Writing the Second World War* (New York: Random House, 2005).
19 As mentioned before, I thank Allen Packwood once more for his indispensable knowledge, advice and help and to Victoria Kelly for permission to write on her father and to reproduce Denis Kelly's words here.
20 Denis Kelly, Letter to Nicholas Bagnall, [no date], The Papers of Denis Kelly, GBR/0014/DEKE, Kelly, Richard Denis Lucien, 1916–1990 (lawyer, assistant to Sir Winston Churchill), Churchill Archives Centre, Churchill College,

Cambridge, Deke 2, 5. I am using the numbers from 138 slides in the file (Deke 2). I have consulted the manuscript in person many times but am using these numbers for those who have access remotely. Thus, I am not using Kelly's own pagination.

21 Kelly, Letter to Nicholas Bagnall, Deke 2, 6.
22 Kelly, "Churchill Memories," 9.
23 Kelly, "Churchill Memories," 9.
24 Kelly, "Churchill Memories," 10.
25 Kelly, "Churchill Memories," 10.
26 Kelly, "Churchill Memories," 10.
27 Kelly, "Churchill Memories," 10.
28 Kelly, "Churchill Memories," 11.
29 Kelly, "Churchill Memories," 11.
30 Kelly, "Churchill Memories," 11.
31 Kelly, "Churchill Memories," 12–13.
32 Kelly, "Churchill Memories," 13.
33 Kelly, "Churchill Memories," 13.
34 Kelly, "Churchill Memories," 13.
35 Kelly, "Churchill Memories," 13–14.
36 For my work on Winston Churchill, I have, over the years, consulted those papers, and I seem to remember them being bound in blue. Part of the description of the papers is as follows at the Cambridge University Library: " The papers of Lord Randolph Henry Spencer-Churchill (1849–1895), Chancellor of the Exchequer and Leader of the House of Commons in Lord Salisbury's 1886 administration, were acquired by the Library in 1994. The collection consists of Lord Randolph's political and personal correspondence between 1880 and his death, and include letters from most leading political figures of the time, Lord Salisbury and Joseph Chamberlain among them." See Papers of Lord Randolph Churchill (Add. MS 9248) https://www.lib.cam.ac.uk/collections/departments/manuscripts-university-archives/significant-archival-collections/papers-lord.
37 Kelly, "Churchill Memories," 14.
38 Kelly, "Churchill Memories," 14.
39 Kelly, "Churchill Memories," 14.
40 Kelly, "Churchill Memories," 15.
41 Kelly, "Churchill Memories," 15.
42 Kelly, "Churchill Memories," 15.
43 Kelly, "Churchill Memories," 16.
44 Kelly, "Churchill Memories," 16.
45 Kelly, "Churchill Memories," 16.
46 Kelly, "Churchill Memories," 17.
47 Kelly, "Churchill Memories," 17.
48 Kelly, "Churchill Memories," 17.
49 Kelly, "Churchill Memories," 18.
50 Kelly, "Churchill Memories," 18.
51 Kelly, "Churchill Memories," 18.
52 Kelly, "Churchill Memories," 19.
53 Kelly, "Churchill Memories," 19.
54 Kelly, "Churchill Memories," 19.
55 Kelly, "Churchill Memories," 20.
56 Kelly, "Churchill Memories," 21.

57 Kelly, "Churchill Memories," 21.
58 Kelly, "Churchill Memories," 22.
59 Kelly, "Churchill Memories," 22.
60 Kelly, "Churchill Memories," 23.
61 Kelly, "Churchill Memories," 24.
62 Kelly, "Churchill Memories," 24.
63 Kelly, "Churchill Memories," 24.
64 Kelly, "Churchill Memories," 25.
65 Kelly, "Churchill Memories," 25.
66 Kelly, "Churchill Memories," 26–27.
67 Kelly, "Churchill Memories," 27.
68 Kelly, "Churchill Memories," 28.
69 Kelly, "Churchill Memories," 30.
70 Kelly, "Churchill Memories," 30.
71 Kelly, "Churchill Memories," 32.
72 Kelly, "Churchill Memories," 32.
73 Kelly, "Churchill Memories," 32–33.
74 Kelly, "Churchill Memories," 33.
75 Kelly, "Churchill Memories," 34.
76 Kelly, "Churchill Memories," 35.
77 Kelly, "Churchill Memories," 36.
78 Kelly, "Churchill Memories," 37–38.
79 Kelly, "Churchill Memories," 39.
80 Kelly, "Churchill Memories," 40.
81 Kelly, "Churchill Memories," 42.
82 Kelly, "Churchill Memories," 48.
83 Kelly, "Churchill Memories," 48.
84 Kelly, "Churchill Memories," 49.
85 Kelly, "Churchill Memories," 50.
86 Kelly, "Churchill Memories," 53.
87 Kelly, "Churchill Memories," 54.
88 Kelly, "Churchill Memories," 54–55.
89 Kelly, "Churchill Memories," 57.
90 Kelly, "Churchill Memories," 58.
91 Kelly, "Churchill Memories," 58.
92 Kelly, "Churchill Memories," 58.
93 Kelly, "Churchill Memories," 59.
94 Kelly, "Churchill Memories," 59.
95 Kelly, "Churchill Memories," 59.
96 Kelly, "Churchill Memories," 59–60.
97 Kelly, "Churchill Memories," 60.
98 Kelly, "Churchill Memories," 60–61.
99 Kelly, "Churchill Memories," 62.
100 Kelly, "Churchill Memories," 63.
101 Kelly, "Churchill Memories," 63–64.
102 Kelly, "Churchill Memories," 64–64.

13 Denis Kelly and Winston Churchill, More Memories and Interpretation

One of my chief emphases in the previous chapter on Kelly and Churchill, this chapter and the book, is on representation and self-representation, so my discussion will continue to focus on that aspect, and I will try to brief. Chapter 4, "Millionaires and the First Stroke," begins dramatically:

> Millionaires I had met were either very dull or very brassy and not people one would ever ask to dinner. The three I encountered with Churchill were of much higher calibre. They were Aristotle Onassis, a self-effacing Jew named Emery Reves, and the famous Lord Beaverbrook.[1]

Kelly provides portraits of these men. He begins the first one thus: "Aristotle Onassis was very quiet, very well-read and, like Churchill, had the gift of expressing profound truths in very simple words."[2] Onassis talks about Napoleon, and the portrait ends with Onasis saying that the truth is simple but takes mental effort to discern. The next portrait starts thus:

> Emery Reves was a Central-European Jew and the exact opposite. He had no charm, like Onassis; no magnetism like Beaverbrook; but he worked like a slave and did more than any other of his publishers to make Churchill's literary fortune.[3]

Kelly quotes Reves about him escaping Hitler's oppression, flinging himself into a sleeping car when stormtroopers entered the railway station in Berlin. Kelly begins the third portrait: "Lord Beaverbrook, on the other hand, was like a brilliant but mischievous – and sometimes malicious – schoolboy."[4] In this portrait, he mentions that Beaverbrook speaks of his Calvinism and capitalism and says that the newspapers of Beaverbrook (a Canadian) were anti-American. Kelly mentions that this was 1949, and that Kelly spoke with Churchill while he was in the bath, was worried about the Russians, felt tired and had a stroke that night.

DOI: 10.4324/9781003651871-14

Taking the narrative back to the day leading up to the stroke, Kelly describes Merle Oberon, the actress, swimming in the Mediterranean at Beaverbrook's private beach and Churchill and Kelly talking and drinking there. Kelly then moves into the evening and night and the next morning:

> When I reached the villa next morning, I discovered it was true. 'Mr. Churchill has had a slight stroke. He wants everyone to carry on as if nothing has happened and he wants you to come up to his bedroom at once.' I did so and found Dr. Roberts – of whom more hereafter -assuring Mrs. Churchill on the telephone to England that it was nothing serious and nothing to worry about. But history had perhaps been changed.[5]

Here, as Kelly notes, is a pivotal moment in Churchill's life and in history. The stroke has private and the public implications. Kelly soon focuses on the medical context where Churchill had the stroke: "Unlike one of his other doctors who is long dead and unworthy of name and wrote a large and financially profitable book about his patient, Dr. Roberts kept his mouth shut."[6] Not only was Roberts discreet, but he also had other qualities that Kelly admired:

> He was a gentle, approachable man who was very kind to us all and was fiercely but quietly determined to wean Churchill from the far too generous supply of sleeping-pills and other drugs the London doctors had been allowing him. Recuperation was slow but perceptible.[7]

As part of recovery, Churchill decided to read William Somerset Maugham's *Of Human Bondage*, which Kelly calls "dismal," and he would not be distracted by the poem by Robert Louis Stevenson that Kelly offered him.[8] According to Kelly, Churchill invited him to dinner in Monte Carlo, where he wanted to be seen to be fit. At dinner, he spoke of the five years out of power: "There was a glint of the old fierce fire in the blue eyes; he had fought the cardiac blow with his war-time slogan 'In Defeat, Defiance'; but to many of us he was never quite the same man again."[9] Churchill, fierce and defiant, but not the same: this is a turning point. Kelly then shifts to the lesson of never going to a fortune-teller and tells of how he consulted one because he wanted to know what would happen after he finished working with Churchill, and her prediction came true that Sir Anthony Eden needed help with his memoirs.[10] Kelly describes how he was vetted for this job and his meeting with Eden and his wife. He describes the unhappiness that the decision to take this job caused him in his "literary life" and, noting that because Eden was in ill health and disappointed, tormented by Suez, Kelly could not serve or help him, so after three months, he resigned.[11] Kelly gives the moral of the story, about seeking out the fortune-teller: "I only tell this tale as a warning to those, like Saul, who may be tempted to dabble in the occult."[12] Moreover, Kelly calls attention to the temporality of his own narrative:

> I devoted myself to the Bar and for fifteen years I wrote no more. And though now approaching seventy, I am still in good health. But all this lay in the future and I returned in the meantime to the radar battles of the second world war – 'The Wizard War' as Churchill called it – and the millionaire-life in which their history was composed.[13]

Kelly is getting ahead of himself, but this episode with the fortune-teller allows him to talk about himself and his future before returning to Churchill and the war. Then Kelly characterizes the millionaires and gives the atmosphere: "It was strange and sterile - cocktail-bars, swimming-pools, servants, expensive food, expensive clothes and expensive cars – which Onassis and Emery Reves thrust modestly into the background and Beaverbrook flourished in one's face."[14] He gives a portrait of these three men and how they hid their work from "the Great Man."[15] Here, by the Mediterranean, Alan Hodge and Churchill (sometimes with Toby the budgerigar in the mix) worked on the literary and historical work at hand for *A History of the English Speaking Peoples*. Besides describing the environment, landscape and atmosphere, Kelly is skilled in representing the three men and their love of Churchill "as a man."[16] Indeed, Kelly shows his love: "Churchill never tyrannized his literary servants. He saw history in strong colours, sombre and dark for Russia and Germany, bright and sparkling for Britain, France and the United States."[17] Kellys says that he and his fellows tried to help Churchill tell the tragedy and observes: "Working with him was like being borne alone by a tidal wave of immense energy. The only problem was to keep one's head above water and not start spluttering and losing one's breath."[18] Once more, Kelly uses image and metaphor to describe, here representing the energy of Churchill as a tidal wave. Kelly also tries to look through the eyes of Churchill: "He saw warmth and colour in everything around him. At the height of the war in Korea I watched him painting a landscape on the plain which stretches from Marrakesh to the Atlas Mountains."[19] Kelly speaks of the art of Churchill's landscape painting and the colourful tapestry of the manuscripts. He is less impressed with Churchill's portrait of Kelly and of one of Churchill's female secretaries: "He never claimed to be a good, still less great, painter."[20] Then Kelly elaborates and sees how important Churchill's art and view of art and life were:

> In a short book called 'Painting as a Pastime' which he wrote between the two world wars and should be compulsory reading for

the millionaires, the workaholics and the unemployed of the end of the twentieth century, he enthusiastically describes how, hard-up, stripped of power and position, with a large family and a vast house to support, and, as it then seemed, no hope of influencing world-events, he taught himself to appreciate and portray the so many beautiful things around us, whether it was a flower in a vase or the farmhouse below Chartwell – 'I painted it in Summer and then decided to cover it with snow' – or building a brick wall round his kitchen-garden – 'Before I started, I was granted membership of the Bricklayers' Union'. (I found his Union card in his archives and he was very proud of it though it was affixed to a letter from a fellow-craftsman saying his plumb-line was askew). He thus proved, in unconscious obedience to the Rule of St. Benedict, that the devils of doubt, despair and introspection can be vanquished by working with one's hands. He never said this in so many words and I don't think he had ever heard of St. Benedict, but both men would have agreed over the centuries that this was the practical solution to Defeat and that the destruction, re-building, re-destruction and rebuilding of Monte Cassino contains a moral we should all dwell upon. 'Painting as a Pastime' is, I believe, the nearest he got to disclosing his philosophy of life.[21]

I quoted this passage at length to show the layers, subtlety and empathy of Kelly's account. Churchill found portraying the beauty round him, and rather than retreat into despair and doubt, he used his hands, choosing creation as opposed to destruction: he built rather than tearing down, and he was positive rather than being given to self-pity.

Kelly is good at giving a sense of those, like him, who surrounded Churchill and his household: "Five remarkable and talented women constituted his permanent secretarial staff. Grace Hamblin, who had served him longer than any of us, presided over the domestic servants and the housekeeping, and attended to the needs of Lady Churchill."[22] Part of his portrait of Hamblin is that she was later the curator of Chartwell and with Lady Churchill worked with the National Trust to transform this house. Kelly also muses about Churchill's home and the last home of Napoleon, and he expresses doubt that these figures would be at ease in the homes if they came back to life. The other assistants Kelly notes: "The other four were responsible for his literary and political exertions and worked in alternate two-or three-day shifts."[23] Kelly provides brief portraits of Elizabeth Gilliatt, Jane Portal, Lettice Marston and "Jo" Sturdee, the last of whom combined them into a happy group that "had to coax, cajole, encourage and serve a temperamental elephant who in some ways seemed never to have quite grown up but was yet much wiser than any of us."[24]

Here in a thumbnail is their relation to Churchill. Kelly says that no one dreamed of discussing Churchill,

> still less of writing a book about him. Perhaps after forty years, they will forgive me for having done so. So many misleading rumours and scribblings have since appeared that I felt I should try to put the record straight, even if my efforts should only amount to a minor footnote to his history.[25]

Here is the nub of the motive for Kelly writing *Churchill Memories*. Forgiveness for breaking the code of silence of those working with Churchill and Kelly's need for correcting the record. Kelly frames his book as setting out the facts, perhaps as an act of loyalty and fealty to Churchill. Furthermore, Kelly signals the passing of a long time, but then says his contribution is a minor part of Churchill's history.

A quotation Kelly came across refers to the twentieth century as that of *The Common Man* because he has suffered the most, and Kelly stresses that Churchill shows compassion for multitudes as he did for the swans at Chartwell.[26] Kelly pushes back against possible stereotypes of Churchill:

> To dismiss him as an imperialist or a war-monger is to miss the whole aim and object of his life. He denounced the Great War as a bloodbath and the Second World War as 'the unnecessary war' which could easily have been avoided with prudence and common sense. He would recite from memory verse after verse of the poems of Siegfried Sassoon on the horrors of trench warfare and spoke with tears in his eyes of the German officer wounded in No Man's Land writing in his diary 'I am lying between the two most civilised nations in Europe'. No one felt more deeply than him that the only desire of the ordinary people of the world is to provide for their families and bring up their children in safety, peace, and reasonable comfort, and providing they have a just Government, they care little what form it takes or what doctrines and theories it may expound.[27]

Kelly calls attention to Churchill's denunciation of war, his recitation of Sassoon about the horror of the trenches, his empathy for a German officer and his feeling for people and their desire to live in peace with their families under a just government. In addition to advocating for Churchill, Kelly defends him against what he considers to be mischaracterizations and misunderstandings.

For Kelly, had Hitler and Stalin followed Churchill's ideal of the well-rounded Renaissance man, the inhabitants of the world would have been happier.[28] Kelly cherishes most the memory of Churchill swimming in his pool at Chartwell and overseeing a race of his grandchildren, playing with them while being concerned with some menace internationally, which causes Kelly to muse about leaders needing to have a happy home and family, calling on Plato and Thomas More to support his musing.[29] Thus, Kelly ends the chapter with this happy and ideal scene.

Chapter 5, "Marrakesh," begins in dramatic fashion, setting the scene with Churchill in a new environment:

> We flew to Marrakesh in perfect weather and a four engine aircraft, and the Capes of St. Vincent and Trafalgar lay clear beneath us. One of the privileges of travelling in a private plane is being invited onto the flight deck and Churchill and I gazed with awe at the promontories where the British naval victories had baulked Napoleon Bonaparte's attempt to conquer the continent of Europe and the Russian Empire.[30]

Kelly mixes past and present, history with a great historical figure, and he describes Churchill and those who surrounded him. Before quoting Churchill's description, Kelly gives his own:

> The air was clammy and oppressive and it was a relief to enter the marble hall of the Hotel Mamounia and contemplate from our balcony-suites the long, historic line of the Atlas Mountains standing like a theatre back-cloth at the far edge of the flat, twenty-mile plain.[31]

Kelly has a talent for atmosphere and setting.

In Marrakesh, they worked and Kelly talks about the contributions of General Pownall and Professor Lindemann and Churchill switched from the Anzio landing to the war in Korea and talked about flying to Tokyo to see for himself. Kelly speculates who put a stop to the trip and what if Churchill had gone in 1952 just as what if he had gone to the European parliament in 1949 (a recurring question in Kelly's account).[32] Kelly and Lindemann accompanied the pilot to go ahead and check for the safety of the runways, Lindemann being quite an expert on flying, part of Kelly's portrait of him.[33] Later, Kelly convinced Churchill that they were flying parallel to the Atlas Mountains and not into them and he tried to distract Churchill with a list of notables they were going to meet:

> among them was a General Turenne, direct descendent of Louis XIV's famous marshal against whom Churchill's own ancestor, the first Duke of Marlborough, had fought at the turn of the seventeenth century. This interested him at once and we landed in calm to a strange reception.[34]

Kelly elaborates – "the French régime had a sense of unreality and one incident distressed me very much" and he tells about three Arab women and speculates the meaning of their presence and of Churchill inspecting the troops. Kelly considers the inspected troops:

> No one told us who they were – I suspected they were same sort of irregulars – or where they were going. "How are things here?" I asked the officer. "Is there any trouble?" "La pacification est complète [sic complète] – the pacification is complete" he answered crisply. Gazing over the desert I had my doubts. The landscape reminded me of the North-West Frontier, but how different from the British Raj in India! There, we learnt the local language, the troops spoke Urdu, we respected and interfered as little as possible with their customs and traditions, we felt part of the country and we and its inhabitants knew that in due course we would leave it – as we did, though with avoidable and tragic bloodshed. Here, one sensed an attempt to convert this wild, alien, Islamic region into a province of France by a rigid and superficial imposition of French civilisation; but I kept these thoughts to myself.[35]

The readers will have to gauge the comparison between the British and French empires. Kelly does admit that there was violence in India when the British later left. Once more, this is a mix of different times in the hindsight. The British may have learned Urdu, but they were not without flaws. Kelly reports on General Pownall's comments on the troops in Marrakesh and his memory of those in Egypt, which reminds Kelly of recent events closer to the time of writing:

> Now, no doubt, it is different. Terrorism, the murder of President Sadat, the horrible wars with Israel and the plentiful supply of arms from the United States and the Soviet Union have bred a different type of soldier – as has happened in Central America. He may be dirty but he is very tough.[36]

The soldiers, according to Kelly, have changed from the ones General Pownall described. Kelly continues:

> One also sensed a feeling of bitterness among the French rulers and one saw it in their faces. They, the French, had been conquered and humiliated by Hitler's Germany. We, the British, had emerged intact and victorious, and they still complained of the inaccurate American bombing of France when military targets were missed and innocent civilians were killed in their homes. Churchill was very sensitive about all this and did his best to cheer them up.[37]

Kelly begins with "One" but it seems that he himself sensed the bitterness. Churchill cheers the French up. Kelly develops the scene, including reported speech, and talks about El Glaoui, the local leader in Marrakesh, whom Churchill knew and all the translation involved at the banquet, including his own role as interpreter at "the most vivid meal I have ever eaten in my life."[38] Then Kelly calls attention to his looking back in these memoirs or memories: "In retrospect the most remarkable feature of the trip apparent was the absence of what is nowadays called Security. Churchill had a single bodyguard – the Scotland Yard detective who, as far as I knew, was unarmed."[39] Kelly amplifies this point at some length. He also muses about security past and present:

> There is much talk nowadays about freedom of information. Much better that there should also be freedom of access to those who govern us and if a few of them get killed as a consequence, why grumble? If you are not ready to put your head on the block at Tower Hill or go, like Abraham Lincoln, unescorted to the theatre, then why be a ruler at all?[40]

Continuing to shift time, Kelly then imagines if Churchill were alive: "But perhaps our civilisation has declined too far for so noble a step and Churchill could now no longer walk freely in the Souks of Marrakesh."[41] Churchill is a great public figure and politician as well as a writer and painter: his fame endures, past and present intertwined.

Kelly returns to the matter of writing: Churchill is always reading, writing, editing:

> Delivery dates rule an author's life and after three weeks we had accumulated a considerable number of drafts which Churchill was impatient to have printed. They were too precious to send by post so I took them to England ahead of the rest of our party.[42]

Kelly describes his time with the British Ambassador and his wife and the flight back on a Dakota of the United States Navy – which reminds Kelly of Burma – from Rabat through Bordeaux to Paris: "The crew disappeared from our hotel but I was too tired to go with them and see the city I had loved as an undergraduate."[43] The description includes a rough landing in the mist north of London, and Kelly creates a typology of Churchill and himself in two places: "It was trying to snow and I was glad that the great man was still in the sunshine of Marrakesh."[44] This is how Kelly ends this chapter before moving to the final movements of the account.

The title page of *The Closing Years*, Chapter 6, includes this handwritten note: "6,100 checked for printer 14/5/85."[45] This chapter begins by getting to the point about the writing:

> It had taken more than seven years to complete the History of the Second World War, and we were sitting alone in the great hall at Chequers over brandy and cigars. It was late at night. The sixth and final volume had at last been finished. The text had been checked and re-checked for misprints, grammar, syntax, mis-statements of fact and for the disclosure of any information which might be of service to the new enemy, Soviet Russia. All that remained was to compose a Preface.[46]

Kelly quotes a passage he wrote and how Churchill revised it.[47] Kelly's narrative gives different detailed accounts of how Churchill worked with his research and writing team. In this chapter, Kelly includes in his account some telling details, for instance:

> I had seen many such instances of his ability to find the right word or the right phrase and the early drafts of his books contain instructive lessons on how to write plain English and indeed how to write a book – from the first embryonic bundle of telegrams and minutes tagged to a brief note into which was slowly woven two or three pages of narrative; ....[48]

Kelly elaborates, and his longer description is valuable; part of the reason I have for years now said that *Churchill Memories* should be published. Moreover, Kelly provides analysis of verbal art, revision, word choice and style in three examples: Churchill's account of his speech to the Parliament in Canada in December 1941; his addition to a long narrative on the German invasion of the Soviet Union of this sentence "Hitler scourged his armies on"; and his choice of the right word in a paragraph of Britain's reconquest of Burma in 1944.[49]

Kelly returns to medical matters and relates that to the writing:

> The second stroke occurred soon after the completion of the History of the Second World War and was more severe than the first, but it led to an unexpected extension of Churchill's literary life and an odd adventure for myself.[50]

Misfortune led to a surprising result. Kelly explains the context clearly and in stages:

> After finishing the four great volumes of 'The Life of Marlborough' in 1934 he had started writing 'A History of the English Speaking Peoples'.

> There were two demanding reasons for this enterprise. 'I knew the war was coming' he told me 'and I knew I should be called to high office, but I also knew that I could only lead the people of our country if I understood how they felt and behaved, so I determined to steep myself in their history'.[51]

Kelly looks back on Churchill looking back on looking ahead to the Second World War. Churchill, as Kelly reports, connects the knowledge of history with political leadership in the present (now past after the death of Churchill). Kelly uses Churchill's voice to help tell his story. He moves to another rationale for Churchill writing this work:

> The second reason was that he had to earn his living by his pen. Few people realise that by the standards of those days and of his own circle he was always hard up – hence the stream of newspaper articles and essays – so hard up, in fact, that when the Second World War exploded into Poland, his publishers refused to pay him unless he finished the book. He was now First Lord of the Admiralty but finish it – in a form – he did.[52]

Writing was the way Churchill made his living. He had to finish his work to be paid, and he did complete it when he was in government. Kelly tells some of the story briefly as it has been recounted elsewhere – "The story has been told by his official biographer" and Kelly says:

> The book was delivered to the publishers early in 1940, his 'advance' was paid, and the manuscript was bundled away into the Chartwell archives while he addressed himself to the business of defeating Hitler. It was never actually printed and sold to the public and he did not touch it again till he had his second stroke in June 1953 when he was once again Prime Minister.[53]

This is background, and Kelly returns to the time of the second stroke with which he starts this chapter. Kelly uses direct speech to set out the sequences of events:

> 'Take up something which will calm your mind' urged his doctors. 'Why not bring "A History of the English Speaking Peoples" up to date?' said Brendan Bracken. A secretary excavated the proofs from the Chartwell cellars and we set to work. We – namely Alan Hodge, editor with Peter Quennell of a distinguished magazine called 'History Today' which had been fostered by Bracken himself – and I followed the pattern which had been set when writing the war memoirs.[54]

The genesis, pattern and method concern Kelly still, and he reiterates this methodology and talks about the missing pages and the solution, being in touch with Churchill's publisher, Cassells. Kelly represents the dramatic moment of the discovery and comments on the significance:

> It was the missing page proofs, the final version delivered early in 1940 which had survived the Blitz, the bombing of Cassell's office near St. Paul's and all the upheavals and removals of a five-year war. It was, as a friend remarked, like discovering the Deutero-Isaiah and to this day, so far as I am aware, it was the only copy remaining in the world.[55]

Kelly describes the great joy for Hodge and Bracken and provides a portrait:

> Alan Hodge was short, untidy, brilliant, and a one-time friend of Robert Graves.... He quivered, like his patron Brendan Bracken, with a restless, unsatisfiable intellect, and knew Michael Foot. His only fault, if one can attribute fault to so loyal and understanding a collaborator, was that he felt too much and too intensely and saw too many answers to a plain question.[56]

Kelly's portrait of Hodge is one of steadfast loyalty: "He was so loyal and modest that he destroyed many of his early drafts for 'The English Speaking Peoples' lest their discovery might diminish Churchill's originality, and he did most of the work of checking the final version."[57] Kelly notes that Hodge is dead and that once he caught his two daughters tobogganing down the stairs on two of Churchill's dispatch boxes – Kelly imagines or recreates the scene with dialogue. Then Kelly provides an overview:

> It took four more years to finish the four volumes of 'A History of the English Speaking Peoples'. Much historical knowledge had accumulated since the original version was hurriedly put together in 1940, many gaps had to be filled and many topics expanded.[58]

Kelly quotes Churchill:

> 'A book' Churchill used to say 'is like an old friend whom you can always go back to' and as the work progressed, so his strength and enthusiasm returned. He fought the second blow with the same courage as at Beaverbrook's villa in 1949 but at the beginning he became tearful and emotional if we touched on any of the great issues of the day.[59]

Kelly links the first and second strokes here and in the wider narrative. He discusses Churchill's health and the reviews of the books. In Kelly's view, this book was not Churchill's best. Churchill thought he was best to write about past leaders and was interested in writing about Napoleon and a history of ancient Rome. Kelly muses about what if Churchill had not lost the General Election of 1945 and Churchill had helped to make the peace and to avoid the position of what Churchill deemed "the sinister shield of the Atom Bomb" standing between humankind and a third world war.[60] Following these speculations, Kelly observes what actually happened: "But old age, infirmity and the intricacies of the Inland Revenue compelled him to retire both from politics and the profession and art of writing, and after finishing 'A History of the English Speaking Peoples' he wrote no more."[61] Kelly adds to that observation the raison d'être for writing his own account of Churchill:

> Many gifted, scholarly and well-informed writers have attempted to describe the Right Honourable Sir Winston Leonard Spencer Churchill, Knight of the Garter, holder of the Order of Merit, Companion of Honour, Fellow of the Royal Society, and ten years as a paid, literary subordinate in his family circle impose humility. In these pages I have tried to avoid both breaches of confidence and of repeating what others have written and told me about him, and I have done my best to stick to what I saw with my own eyes and heard with my own ears between 1947 and 1959.[62]

Kelly portrays himself as a literary assistant and an eyewitness who has been humble and discreet and has limited himself to his own witnessing to avoid repetition. He then says he has tried to avoid hearsay and hopes his work will be humble like the essay of Churchill's daughter, Sara, that Kelly praises. For Kelly, the great are people no happier than others and Churchill, a father and husband, "shouldered these universal afflictions without murmurings or complaint."[63] Kelly tells of how one evening Churchill and he had discussed what their lives would have been like had they been women. That leads Kelly to recount Maupassant's story about a French prostitute called "Boule de Suife" or "Ball of Lard." Churchill in turn asks if Kelly had read "Le Roman D'un Jeune Homme Pauvre?" Both stories involve discrimination and honour.[64] Kelly turns to Churchill's love of, and respect for, food. Kelly muses on the beneficial mixtures of certain alcohols and foods and tells of his dining at the English College in Rome and how the infirmarian there thought of the benefit of wine and port with meals. The many gifts Churchill received from round the world Kelly chronicles, including "a large, beautiful and delicate painting of Mount Fujiyama which stood on an easel in his studio and was presented by the Japanese within a few years of the casting of the atomic bombs."[65] The poignancy of this gift, which Kelly calls "most remarkable," is implied but not remarked on.[66] Kelly has a gift for description and telling details. For instance, he muses on Chartwell then and now and the cast of characters who watched films with Churchill:

> The dining room which contained these conversations has now been turned into a museum and I regret that Chartwell was not allowed to remain as it was when he died. What is now the breakfast room was once a cinema and after dinner on Saturday night a film was displayed to which all were invited from the most distinguished guest to a gnarled and respected old man named MR. KERN who worked on the farm and collected Churchill's cigar-stubs and ground them down to smoke for him in his pipe.[67]

Kelly describes the projectionist, Mr Shaw and his wife, as socialists and how Sir Alexander Korda had a hard time finding films for Churchill who liked films in which men were heroic and women beautiful and true and in which right triumphed over wrong. Kelly includes the direct speech of Churchill's words and tells of Churchill's favourite movies and how he would give his verdict when the film was done. The day would end at 2 a.m., and the film was between the work of the afternoon and of the night, work on subjects such as the American struggle against the Japanese in the Pacific.[68] For Kelly, Churchill was as wise as he was human: "Now, in spite of his efforts and warnings in the wilderness years, it had all happened again."[69] Churchill had experienced the First World War and warned about the coming war and later tried to avert a third catastrophe. The human element Kelly also shows in Sutherland's portrait of Churchill that Parliament had given him in 1954 to mark his eightieth birthday:

> He was as unself-conscious as a child and the same person whether he was in the House of Commons, at a Garden Party at Blenheim, or in the swimming-pool at Chartwell, but touchy about his personal dignity. 'I never looked like that' he grumbled at Graham Sutherland's portrait. 'I once saw you look like that' answered Lady Churchill. 'It was during the war, after the fall of Crete.'[70]

In a sense, this is a kind of small ekphrasis, a description of a picture, bridging the visual and verbal, a rhetorical device imaging the tale behind the visual art and going back at least to Homer. Kelly elaborates on this portrait within a portrait, this metaportrait, to muse on Churchill and the great leaders not being able to "show fear, doubt, despair, despondency or a wavering determination."[71] Next Kelly speculates on Churchill sensing his loss of power at eighty being at the heart of his complaint about the portrait, and Kelly quotes James Gunn, a portrait painter, on the mystery of this kind of painting as a glimpse at what the painter thinks is the truth about, or reality of, the sitter. Kelly then discusses Andre de Moller and his portrait of Margaret Thatcher and how de Moller thought a picture can take on a life of its own beyond the creator and his account of an experience of painting a host in Ireland that was like a picture of Dorion Gray.[72]

Kelly likes comparisons and contrasts to sharpen his portrait of Churchill. The art of painting he compares with the art of writing – the difficulty of portraying someone else. Kelly also brings together comparisons in his own life in history and in law:

> Rayner Goddard, Lord Chief Justice of England from 1946 to 1958 was, of all the men I have encountered, the one who most resembled Winston Churchill. Both were rough, tough and well-read, with a relentless appetite for work, food and wine; a rustic commonsense; a sharp wit; an innate, unquestioning patriotism; and the ability to make considerable mistakes. But both men had the honesty and courage to admit, if you stood up to them, that they were wrong. Like Churchill, he was intolerant of sloppy English and ill-digested, clumsily-expressed ideas and it was prudent before addressing him to clear your mind and refine your thoughts by putting them on paper beforehand and boiling them down into a few, blunt, simple words.[73]

Here, in this comparison, of which this passage is but a part, Kelly portrays two characters alike, Goddard and Churchill, who had the ability to admit mistakes and who demanded simple, clear English. These men were fair-minded. At the end of the comparison, Kelly shows another reason for comparing them:

> I never met Goddard socially but I learnt afterwards, having toiled over Churchill's chapter in 'A History of the English Speaking Peoples' on the English beginnings of the Common Law, that Goddard had vetted it and that Churchill had pronounced him 'a splendid old man'.[74]

What might seem a slight autobiographical indulgence makes sense as Churchill calls on Goddard and compliments him, and Kelly and Goddard meet virtually in reading Churchill on the origins of the Common Law. Here court and writing meet – the two sides of Kelly are also the two sides of Goddard, who is like Churchill. Perhaps Kelly aspires to being like Churchill and Goddard in the matter of writing.

As Kelly approaches the end of his book, I approach the end of this chapter. In my own modest way, I have tried to do for Kelly what he did for Churchill in his memories, to shed a little light on him. When researching Churchill, I came across many figures, including Kelly. One is aware in life and in the archive of what Edmund Spenser described as endless work. Perhaps I have given too much space to Kelly even if he deserves more because had someone published "Churchill Memories," I would not have felt so bound to give such a detailed sense of this work. Perhaps someday the book will get into print. In the meantime, here is a portrait of Kelly as he gives a portrait of Churchill.

Kelly closes Chapter 6 with a few more interesting elements. He represents a farewell:

> We said goodbye in the bedroom of his London home at Hyde Park Gate. He was dressing for lunch – he usually worked in bed till midday – and after pulling on his shirt he looked at me very steadily and answered: 'Not goodbye but au revoir'. Except for two brief occasions I never saw him again till his funeral at St. Paul's Cathedral.[75]

Kelly then shifts to the morning of Churchill's funeral in January 1965, which I remember, in my ninth year, seeing on television, and Kelly was reminded then of the words of Winston Churchill when William Deakin, Kelly and he were working on a difficult chapter and Churchill called them "boys," although they were about forty, and said how they must "hate" this work, adding "Mind you have a jolly good time at my funeral."[76] Kelly gives voice to the voice of Churchill. A great man with a human touch.

Then Kelly contemplates Churchill and age:

> Old age is a mysterious condition. I saw little of him during the last years of his life but he reminded me of an enormous bonfire. You think it is black and dead but suddenly, by some freak of chemistry or a chance gust of wind, it suddenly expels a brilliant spark which soars into the sky, glitters and glows, and then vanishes as swiftly as it came.[77]

This is a beautiful and poetic description. Kelly is a writer as well as a historian and lawyer. He speaks of Churchill's idea of rockets transporting mail and Britain and the United States having a common currency. Kelly also says of Churchill: "He had a clear policy about Death."[78] The word "policy" is both curious and arresting here in regard to death. Kelly quotes Churchill on preferring the Almighty giving him "eternal sleep" than "the government of the Universe."[79] Like many of us, Churchill, as Kelly – quoting from Marcel Proust – says, grew troubled in the long dark nights as he grew older, by what was done or should have been done and sought the oblivion of sleeping pills. This haunting haunted Kelly:

> I wondered, though I was not present, if his restless brain still disturbed him in the days and nights when he lay unconscious at 28 Hyde Park

Gate in the bedroom I knew so well during the final massive stroke which preceded his death.[80]

This is perceptive of Kelly, as it seems that the brain works and responds in people who are unconscious, in comas and under anaesthetic. Then Kelly does rely on the report that Churchill's face was "smooth and untroubled" and that he had told a friend that he had wanted to die on the anniversary of the death of his father, which he did.[81] Kelly tells of a joke and a legend in which Churchill appeared on his doorstep and announced the death of his doctor, Lord Moran. Even in death, Churchill was full of wit and was legendary. Kelly explains: "The joke was current at the time and is now well-known, but wherever he is, the Great Man would appreciate the legend."[82] And thus ends Chapter 6.

On the title page, seventh and final chapter, "Postscript," Kelly has written in blue ink "Ready for printer 14/5/85."[83] At the end of the one-page chapter, he signs "Denis Kelly. 22 May 1985" in black ink.[84] Kelly begins this one-page "Postscript" with these words: "I kept no diary of my ten years' work with Churchill and any papers I possessed were long ago returned to his archives. This was deliberate. We had an unspoken agreement that what passed between us was sacred."[85] Kelly is clearly worried about breaking that trust. He quotes a French Prime Minister speaking to a former French President in which he alludes to thirty pieces of silver, the amount paid to Judas Iscariot to betray Jesus. Kelly wants others to know he is not Judas.

Here is Kelly's final defence in the final paragraph of the book:

> Forty years have now gone by and I hope that these pages have not broken the trust he honoured me with. An official biographer is hampered by no such constrictions. His duty is to expose the truth, however unpalatable or disconcerting, but a literary-devil cum house-guest remains bound by the vow of confidentiality. I can only hope that these rambling reminiscences may nourish a deeper understanding and appreciation of a man whose courage, talent, faults and determination saved the civilisation of the twentieth century, and that he and his descendants will tolerate this attempt to provide, as did Sarah Churchill, another thread in the tapestry of his life.[86]

The revisions are telling. In black ink, Kelly deletes "reflections may contribute to" and changes the phrase to "rambling reminiscences may nourish" while inserting "faults" in red ink and adding "of his life" in black ink. Kelly is facing the Churchill family and hopes they will "tolerate" his "attempt" as much as they would a family member like Sarah Kelly broke his silence forty years later and seems, with Proust and Churchill, to look into the dark night and, perhaps with Saint John of the Cross, the dark night of the soul. In Kelly, there seems to have been a tension between upholding the silent agreement of not telling and his desire, in his own later years, to tell the story and ask for tolerance and perhaps forgiveness. In Kelly's old age, perhaps the brilliant spark arose from the dark, from his bonfire. Kelly, while fair to Churchill, someone he clearly admired, may have had an "imp" or mischievous child within him, something implanted in him to go back to the Greek, Latin and Old English and other roots of the word. Perhaps we all have imps within us, and it is not always easy to keep them at bay. Some time ago I wrote to Allen Packwood, the Director of the Churchill Archives Centre, about this aspect of Kelly, whose writing I admire in many ways as I hope my writing here about him shows. Here is an excerpt of what I wrote to Allen:

> Late in life, when undergoing treatment for cancer, Kelly seems to think Churchill married the wrong person. He alludes to Freud but seems to focus more on Hitler as a homosexual than on the sexuality of Stalin and Churchill. In another letter, a confidential one of three pages to Richard M. Langworth of the International Churchill Society dated 2 September 1988, Kelly shows his other side, speaking of how hard it is to place his autobiography of his childhood: 'Sandwiched, however, between my rambling reminiscences is a section of my ten years' service – (1947–1957) – as one of the literary assistants to Sir Winston Churchill, ten years in which i physically sAt [as typed in the original] beside him at Chartwell and elsewhere and acted as a post-boy, "re-write man", and literary Man Friday when he composed his six volumes about the Second World War and his four volumes on a History of English Speaking Peoples. I also catalogued his archives.' Later on the page, Kelly writes: 'I was a paid and confidential employee and in the well-known words of a French Prime Minister, I have no intention of betraying my Master for 30 pieces of silver.' So, in a sense, in one letter Kelly gossips about Churchill and his wife but says he will not betray Churchill in this letter earlier in the month.[87]

And so I do not, cannot, should not cast the first stone. I am all too human. Despite quirks, which I imagine we all have, Kelly makes an important contribution to the study of Churchill. Kelly's autobiography and biography, his self-representation and representation are worth careful consideration. I hope that I have done some of that work here. Just as Boswell was to Johnson, so too was Kelly to Churchill. The circumstances were different, but both Boswell and Kelly have some insight into the great men they accompanied and wrote about. There were and are others who continue to interpret Samuel Johnson and Winston Churchill, but Boswell and Kelly help to understand the respective figures. Boswell wrote more, was published and is better known, but Kelly, although unpublished in his remembrance, deserves to be better known. That is no slight compliment to Kelly to whose memory I mean no slight. I am thankful for his memories of Winston Leonard Spencer Churchill, and I wanted to pass on that thanks to others. There is more to know about Kelly, his other writings, his interviews, but here for now is what I consider to be his greatest contribution as a writer and historian.

Notes

1 Denis Kelly, *Churchill Memories* [or "Churchill Memories" as it is an unpublished manuscript], Churchill Archives Centre, Deke 2, p. 66. The numbers correspond to the slide number in Deke 2 and not to the pagination that Kelly uses.
2 Kelly, "Churchill Memories," 66.
3 Kelly, "Churchill Memories," 67.
4 Kelly, "Churchill Memories," 68.
5 Kelly, "Churchill Memories," 71.
6 Kelly, "Churchill Memories," 72.
7 Kelly, "Churchill Memories," 72.
8 Kelly, "Churchill Memories," 72–73.
9 Kelly, "Churchill Memories," 73.
10 Kelly, "Churchill Memories," 74–75.
11 Kelly, "Churchill Memories," 77.
12 Kelly, "Churchill Memories," 78.
13 Kelly, "Churchill Memories," 78.
14 Kelly, "Churchill Memories," 79.
15 Kelly, "Churchill Memories," 79.
16 Kelly, "Churchill Memories," 80.
17 Kelly, "Churchill Memories," 81.
18 Kelly, "Churchill Memories," 82.
19 Kelly, "Churchill Memories," 82.
20 Kelly, "Churchill Memories," 83.
21 Kelly, "Churchill Memories," 84.
22 Kelly, "Churchill Memories," 85.
23 Kelly, "Churchill Memories," 85.
24 Kelly, "Churchill Memories," 86.
25 Kelly, "Churchill Memories," 86.
26 Kelly, "Churchill Memories," 87.
27 Kelly, "Churchill Memories," 87.
28 Kelly, "Churchill Memories," 88.
29 Kelly, "Churchill Memories," 89.
30 Kelly, "Churchill Memories," 91.
31 Kelly, "Churchill Memories," 91.
32 Kelly, "Churchill Memories," 92–95.
33 Kelly, "Churchill Memories," 96–97.

34 Kelly, "Churchill Memories," 98.
35 Kelly, "Churchill Memories," 99–100.
36 Kelly, "Churchill Memories," 100–101.
37 Kelly, "Churchill Memories," 101.
38 Kelly, "Churchill Memories," 102.
39 Kelly, "Churchill Memories," 105.
40 Kelly, "Churchill Memories," 106.
41 Kelly, "Churchill Memories," 106.
42 Kelly, "Churchill Memories," 107.
43 Kelly, "Churchill Memories," 107.
44 Kelly, "Churchill Memories," 108.
45 Kelly, "Churchill Memories," 109.
46 Kelly, "Churchill Memories," 110.
47 Kelly, "Churchill Memories," 110–111.
48 Kelly, "Churchill Memories," 112.
49 Kelly, "Churchill Memories," 112–114.
50 Kelly, "Churchill Memories," 115.
51 Kelly, "Churchill Memories," 115.
52 Kelly, "Churchill Memories," 115.
53 Kelly, "Churchill Memories," 115.
54 Kelly, "Churchill Memories," 115–116.
55 Kelly, "Churchill Memories," 117.
56 Kelly, "Churchill Memories," 118.
57 Kelly, "Churchill Memories," 118–119.
58 Kelly, "Churchill Memories," 120.
59 Kelly, "Churchill Memories," 120.
60 Kelly, "Churchill Memories," 121.
61 Kelly, "Churchill Memories," 121.
62 Kelly, "Churchill Memories," 122.
63 Kelly, "Churchill Memories," 122.
64 Kelly, "Churchill Memories," 122–123.
65 Kelly, "Churchill Memories," 125, see p. 124.
66 Kelly, "Churchill Memories," 125.
67 Kelly, "Churchill Memories," 125.
68 Kelly, "Churchill Memories," 125–126.
69 Kelly, "Churchill Memories," 127.
70 Kelly, "Churchill Memories," 128.
71 Kelly, "Churchill Memories," 128.
72 Kelly, "Churchill Memories," 129–130.
73 Kelly, "Churchill Memories," 131.
74 Kelly, "Churchill Memories," 132.
75 Kelly, "Churchill Memories," 133.
76 Kelly, "Churchill Memories," 133.
77 Kelly, "Churchill Memories," 134.
78 Kelly, "Churchill Memories," 135.
79 Kelly, "Churchill Memories," 135.
80 Kelly, "Churchill Memories," 135.
81 Kelly, "Churchill Memories," 135, see p. 136.
82 Kelly, "Churchill Memories," 136.
83 Kelly, "Churchill Memories," 137.
84 Kelly, "Churchill Memories," 138.

85 Kelly, "Churchill Memories," 138.
86 Kelly, "Churchill Memories," 138.
87 My email to Allen Packwood is part of my notes from the Churchill Archives Centre 24–28 April 2024. Allen Packwood has been generous in his discussions of Churchill, Kelly and others for almost two decades now. The Kelly letters I allude to are in Deke 1 at the Churchill Archives Centre, a marvellous treasure for Britain and the wider world.

14 Ecology, Natural History and Ecospatiality in American Literature

Since the Second World War, industrialization intensified and so did the interest in ecology. The tension between development and the destruction or alteration of the natural world has become a theme in literature but has long been a concern in science and culture. In the 1980s, when ecological matters were pressing and much debated, Denis Kelly was looking back to his time in the 1940s and 1950s working with Winston Churchill. The body of my book began with Charles Darwin, natural history and natural selection and here, I end the body of the volume with a return to biology and nature, the question of ecology and its relation to literature and culture.

Nature, history, literature, mathematics and science were interests I had from my earliest years, so I am interested in the recent works of Juliana Chow, Alexander Menrisky and Lowell Wyse on natural history, ecology and ecospatiality in American literature respectively.[1] At the Harvard University Herbaria, I have the privilege to be associated with a place with a long tradition of distinction, including someone I have known and admire for decades, Donald H. Pfister. He studies fungi, especially ascomycetes and employs molecular, morphological and life history information to comprehend the connections amongst these fungi as well as their activities in nature. Pfister and I are interested in the contributions of his predecessors, such as Asa Gray, at the Herbaria and Harvard and to Boston, the United States and the wider world.[2]

Science, history, literature and culture all meet in this place. For instance, Asa Gray, who started the Harvard University Herbaria, corresponded with Charles Darwin and disseminated his ideas in the United States. Janet Browne sets this out well: "When Charles Darwin died on 19 April 1882, Asa Gray lost not only a longstanding friend but also his closest intellectual partner in the development and dissemination of Darwinism in the United States."[3] She includes an illustration of a manuscript of a letter from Gray to Darwin in December 1859, the month after the publication of *Origin of the Species*.[4] In 1959, Hunter Dupree had written at some length on this

DOI: 10.4324/9781003651871-15

connection between Gray and Darwin.[5] Part of this friendship was based on writing and getting word out. This is a dimension that interests me, and American literature represents ecology and natural history, a space for biology in society and culture.

Biological sciences in the Atlantic world were connected, and the connection between Britain and the United States helped to move that subject in new directions, including natural selection. Stephen Jay Gould wrote essays, gathered in *Reflections in Natural History*, which is, as Jaboury Ghazoul notes, "a reflection on the interplay of ecology and evolution," and Ghazoul himself maintains: "Ecology only really makes sense in the light of evolutionary theory."[6] Charles Elton considers ecology to be "a new name for a very old subject. It simply means scientific natural history,"[7] and some basic ecological concepts of ecology can be seen in writing on natural history going back to the ancient Greece.[8] Ghazoul views ecology as moving from being a science to a worldview[9] and, more generally, he examines ecology in culture, arguing that insights in ecological science come to be transferred to political and moral fields, asking whether ecology as a science has allowed for the development of "an ecological conscience" that will help to withstand challenges in the environment, outlining movements, such as Gaia Theory (the earth as a living organism), Deep Ecology (that is not human centred), Cultural Ecology (the ability of the environment to sustain culture), Sacred Ecology (a way to interact with nature).[10] For Ghazoul, the ecologist has gone from binoculars, a tape measure and a notebook to Big Data and some new technologies and questions in ecology; once bearing little on politics, now it means that ecologists are involved in areas that often involve conflict and politics "conservation, land management, and resource use" during "the Anthropocene epoch" that involves "biodiversity losses, climate change, elevated atmospheric carbon, ocean acidification, nitrogen deposition, the spread of invasive species, deforestation, and soil erosion."[11] Education of young children in natural history may be a way to a sustainable society.[12] Why I have chosen these three works, Chow, Menrisky and Wyse, all of whom end their work with hope as Ghazoul does, is that they are on natural history or ecology, which are one and the same, different terms for the same subject. Each of these authors employs different key terms in their titles to explore this field as expressed in American literature: natural history, ecology and ecospatiality.

"Natural history," an ancient term, is the one Juliana Chow prefers in her study. Before discussing that work, I turn to the etymology and meaning of the term. The origin in English is a compound based on the classical Latin *nātūrālis historia*, for instance, in the title of a text by Pliny the Elder, itself after the ancient Greek *φυσικὴ ἱστορία* from Aristotle, a comparative in Middle French and the French *histoire naturelle*, a work on the natural world during the middle of the sixteenth century, the branch of knowledge of all natural objects (1765).[13] The definition in the *Oxford English Dictionary*, which quotes J. Fewterer's translation of U. Pinder *Myrrour Christes Passion* (1534) f. iiii[v] "as Plinius sayth in his naturall historie," shows the range of, and change in, "natural history":

> A work dealing with the properties of natural objects, plants, or animals; a systematic account based on observation rather than experiment. Now chiefly in the titles of books dealing with the wildlife of a particular region or the biology of particular organisms.[14]

Chow takes up that subject in her recent book.

Chow's *Nineteenth-Century American Literature and the Discourse of Natural History* (2022) explores literary representations of natural history in terms of feminism in relation to science and critical race theory and discusses diaspora and "biogeography" while examining writers of colour like Simon Pokagon and James McCune Smith.[15] Moreover, Chow analyses literary works depicting questions of survival and fragility in the environment in connection with Darwinian evolution and the westward colonial expansion of settlers in America. To begin, Chow provides a close reading and says: "In 'missing All,' Emily Dickinson's poetic eye misses the sweep of the sun's course across the sky that extends and engulfs from light into dark."[16] Poetry becomes a lens for natural history. Chow ends the reading with the observation about the speaker in Dickinson's lyric: "Her diminished view of her world and her work—on axial rotations that are at once celestial and quotidian, geographical and bio-logical—still has the capacity to apprehend extinction as a partial but palpable loss."[17] From Dickinson, Chow turns to the German naturalist, Alexander von Humboldt, whose "geographical approach especially influenced popular and scientific views of nature and the development of ecological thinking in nineteenth-century America."[18] Dickinson's image of the partial and of loss becomes a metaphor for an aspect of natural history in her time and for Chow's book: Chow asks, "How does a partial view—a partial biogeography or a partial ecology—modify senses of environmental and biological relations? How does it change senses of species extinction or survival, of species development or transformation?"[19] This partiality or diminishment becomes a central concern of Chow's volume. She observes:

> In an era where the scale of human-induced environmental degradation encompasses the entire planet—from climate change to the alteration of the geological record now called the Anthropocene—it may seem perverse to turn to a perspective of diminishment.[20]

This is a critical time in natural history or ecology. Furthermore, Chow describes an alternative way:

> This study retreats to the grounds of Americanist ecocriticism and the experiential conundrum of subject–object relations to glean how biogeographical materialisms of diminishment already offered alternate figures of survival at the very moment when an American environmentalism based on appropriation, preservation, and recovery began to be articulated.[21]

Diminishment is another figure in an interest in the environment. There is a historical element: "Settler colonialism's geographical and optical expansions contributed to knowledge of ecological relations even as it depleted natural resources and threatened species loss."[22] This is a double movement. Chow comes to the nomenclature and its history: "Naming 'ecology' in the mid-nineteenth century, Ernst Haeckel emphasized the science as a comprehensive and systemic sense of nature investigating."[23]

The structure of the book builds the argument from the casualties of natural history through sketching American species, the particularity of disjunction, race and diaspora to dispersion and writing a natural history of casualties. In Chapter 1, Chow provides a framework of sight to organize her critique that connects literary and aesthetic observation to the idea of species diminishing in the nineteenth century, discussing a diminished seeing in sketches of nature by John James Audubon, Susan Fenimore Cooper and Simon Pokagon.[24] In Chapter 2 (69–97), Chapter 3 (99–141) and Chapter 4 (142–182), Chow employs case studies of Emily Dickinson, James McCune Smith and Henry David Thoreau and their scientific context, setting out each writer's readings and possible links with scientists such as Charles Darwin, Charles Lyell, Louis Agassiz, Asa Gray and James Dwight Dana. For example, Chow says about Asa Gray: "In attributing floristic relations between eastern North America and Japan, Gray famously aligned himself with Darwin and the possibility of species transmutation qualified by the crucial proviso of geographical conditions and consequences."[25] Darwin, Gray and Dickinson all consider "lines."[26] Chow examines the effects of linking biogeographical investigations of species and their "development and dispersal to ethnological theories" of race, and she continues to pursue "a natural history of diminishment by a retreat to a perspective peripheral yet proximate to the heart of settler-colonial power, this chapter focuses on how the New York African American physician and organizer James McCune Smith."[27] Human notions of race and the natural world Chow explores culture and science being brought together in her analysis. She concludes:

> Against the segregation of races and species, the microscopic zoophyte quietly building its coral reef under the sea's surface is a strikingly alien and composite life form brought into relation to human civilization that materializes the union that difference—as diminishment—can make.[28]

Chow's case studies have different emphases, concentrating on Emily Dickinson and the possibility of species as figures (disjunction, metaphor) and the "gaze" of biogeography, focusing on James McCune Smith's widening of geopolitical consequences of the dispersal of species, exploring how Henry David Thoreau informs the sketch illuminates "ecological relations" and the resultant "historiography of nature."[29] Discussing Thoreau's notes on Fire Island, she quotes the last few lines, including: "I saw many leaves of a large unbound Latin Book–scattered over the beach a mile from the wreck. The Universal Anatomy of Paul Mascagnius Pisa1826"[30] and reproduces on the next page "Henry David Thoreau Field Notes from Fire Island (Figure 4.3 MS Am 3032 Leaf 9 verso, Houghton Library, Harvard University."[31] The Herbaria and Houghton Library at Harvard house some of the evidence on which Chow bases her observations. She relates text and context, culture and nature, pages washed up from a shipwreck on the waves and land:

> Thoreau stopped to examine them and scrupulously gleaned reference information from a title page. The corpus is at once text and tissue; unbound and eroding; a small clearing of history scattered among the rocks, a material history, a partial history.[32]

In the "Afterword," Figure 5.1 on page 184, Chow has a figure of "Asclepias cornute specimen from the Henry David Thoreau Herbarium (Gray Herbarium, Harvard University)."[33] Chow views the evidence thus: "The specimens reveal something of Thoreau's plant collecting practice—he must have wrapped them in newspaper to carry them during his walks."[34] This kind of analysis, concrete and practical, is welcome.

Chow concludes her study with an interesting view, creating a typology between a contemporary film and nineteenth-century natural history: Lynnette Wallworth's film *Coral Rekindling Venus* represents "reef life and coral polyps swaying, opening and closing, gathering nutrients, reproducing, and fluorescing in the format of a planetarium show that typically projects constellations and other celestial phenomena."[35] The role of metaphor in language, poetry, scholarship and science is an old debate, and here Chow's poetic description, her use of analogy (Thomas Aquinas discussed that), of a film representing a coral reef, that is, nature. Language and mathematics, culture and biological sciences meet in interpretations of ecology. Chow evokes facing coral and ends with empathy: personification is a type of metaphorical joining with nature, a kind of animistic yearning, perhaps an ancient human feeling in nature, when she says of this film of coral: "it asks us to feel the enigmatic life of a coral flower as a singular discrete trope," and she continues:

> This turning perhaps produces both centripetal and centrifugal forces, and though we cannot know whether these will be enough for species such as coral to adapt to climate change and other environmental and biological transformations, it is precisely those fleeting acts of living and dying, gathering and dispersing, gleaning and letting go, that might allow for survival.[36]

An etymological gloss will help an interpretation of Chow's interpretation here: "(iii) classical Latin *tropa* solstice (attested in an inscription; also mid 16th cent. in post-classical Latin), and its etymon (iv) ancient Greek *τροπή* turn, turning, solstice < an ablaut variant (*o*-grade) of the base of *τρέπειν* to turn."[37] She plays on words in her synecdochic and metonymic use of metaphor and of a trope, Latin *tropus*; Greek *τρόπος*, a figure of speech, but the Greek means a turn, and the Oxford English Dictionary helps to explain her move from trope to turn to solstice.[38] By analogy, the film represents and embodies in the coral a turn to a solstice that arouses human feelings that might permit "survival." This is a hope.

In *Wild Abandon* (2021), Alexander Menrisky examines the identity politics of ecology (IPE) in American literature. "Ecology" came into English from the German *Oecologie* (now *Ökologie*) in E. Haeckel's *Generelle Morphologie* (1866) (II. 286), which comes from the ancient Greek *οἶκος*, house for dwelling, and is related to the French *écologie* (*c*1874) and means the following: "The branch of biology that deals with the relationships between living organisms and their environment. Also: the relationships themselves, esp. those *of* a specified organism."[39] For Menrisky, the American wilderness narrative separates nature from culture and continues even with the development of ecological science, which stresses the connection and not the separation of the natural and the cultural. He explores how in the twentieth century ecology interacts with the politics of authenticity to alter that narrative and argues that in the 1960s and 1970s, ecology became more political and that often environmentalists used the ecological with psychoanalysis and identity politics, so that ecology was a position of identity that culture expressed through social difference and selfhood, repressed. Moreover, Menrisky interprets works by Edward Abbey, Simon Ortiz, Toni Morrison, Jon Krakauer and others, including Margaret

Atwood, thereby interpreting American as North American. Menrisky examines how these texts embody the relations and even tensions among ecology, identity, authenticity and provide alternative representations and histories.

There are many dimensions to the ecological in American literature. For instance, Menrisky says: "Twentieth and twenty-first century literature teems with scenes of dissolution, moments at which a text appears to erase individuals' sense of self-identity in natural environs."[40] This is a key tension between nature and the human. The author is explicit in his goal: "*Wild Abandon* is about the cultural narratives that emerged when political radicals of the 1960s and 1970s joined ecology with psychoanalysis."[41] Culture is Menrisky's concern and the role of story or narration. More specifically, he says: "This book will argue that an interaction between ecological science and midcentury social theory gave shape to what I term an *identity politics of ecology* (IPE)."[42] This sense of ecology focuses on politics and the debate over identity that was especially intense during the 1960s and 1970s. Menrisky argues that the IPE identifies "holistically with the ecosystem writ large."[43] The connection between the science of ecology (describing "material networks and interactions") and cultural practices (mediating and interpreting) is one that Menrisky stresses.[44] *Wild Abandon* examines an IPE across various postwar institutions and environmental traditions and analyses how literary works represent conflicts between the idea of ecological authenticity and other notions of identity by staging the dissolution of the ego, thereby uncovering the "representational and political dilemmas" of the narrative of the IPE.[45] Furthermore, Menrisky elaborates on the method: "The principal question that guides this book's analysis is: *What cultural work is dissolution performing at any given moment, and why?*[46] The literary texts support or question the story of the authenticity of ecology."[47]

In Chapter 1, Menrisky maintains that Edward Abbey shows how radicals unintentionally undermined their own project of self-fulfilment and that, in *Desert Solitaire*, Abbey represented subjective confusion in his wilderness alternative. Menrisky argues that Abbey and other nature writers of the 1960s represented a tension between self and system that is characteristic of literary works on ecology in this period.[48] For instance, Menrisky says: "In *Desert Solitaire* he engages two of the New Left's most consistent characteristics: its policy of direct governance and its emphasis on a politics of authenticity."[49] Chapter 2 discusses how those in the American counterculture and the New Age called on ecological authenticity to stress that dissolution gives access to an identity originating with the ecosystem. Menrisky asserts that Simon Ortiz's *Woven Stone* (1992) reveals how countercultural appeals to ecological authenticity misinterpret and misappropriate perspectives and shows how language and narrative make ecological affiliations and enhance them.[50] In *Woven Stone,* which collects three poetry collections, Ortiz "articulates the sort of connection to the earth often attributed to Native cultures in terms of its narrative basis, rather than psychic primacy" and "interrogates settler projections of authenticity onto those same cultures."[51] Indigenous and settler cultures meet in the interpretation of ecology. Chapter 3 explores how appeals to ecological authenticity might be identity politics, relates them to identity movements that arose with modern American environmentalism and examines Toni Morrison's *Song of Solomon* (1977).[52] Menrisky observes: "*Song of Solomon* offers insight into these developments for its remarkable juxtaposition of two representations of authenticity: the racially particular and the ecologically universal."[53] Race and ecology connect here. In Chapter 4, Menrisky analyses how the IPE affected different forms of identity politics and provides an interpretation of *Surfacing* (1972) along with Margaret Atwood's later work in which he sees a rhetorical continuity between the essentialist position in nature feminism and the position in new-materialist writing.[54] Menrisky argues that "*Surfacing* reads far more compellingly as a narrative of self-undiscovery"[55] and "takes an anti-essentialist stance, a position visible in other texts of Atwood's oeuvre."[56] What does ecology or nature mean for the sense of the human, to its self-discovery or ego? Does it dissolve ego and undiscover or allow for a deeper connection within a human and between humans? Chapter 5 looks at *Into the Wild* as a cautionary tale and concludes with the view that an IPE is impossible and undesirable politically.[57] According to Menrisky,

> Jon Krakauer published *Into the Wild* (1996), a reconstructed account of [Christopher] McCandless's cross-country travels, over the course of which the man fashioned himself as a self-renouncing ascetic icon: the 'Supertramp,' an alter ego styled as a paragon of authentic selfhood, unfettered by social custom, continuous with his environment.[58]

What is an authentic relation with self and environment in the context of ecology? Menrisky also stresses how reading practices affect environmentalist aims positively or negatively.[59] Menrisky concludes *Wild Abandon* thus: "Wilderness continues to loom large in political and literary culture. What do we learn when we read that narrative today? Nothing about ecology's authenticity, but a lot about what people think about it."[60] Wilderness remains central in American culture, politics and literature, and Menrisky is interested in the interpretation of ecology but does not consider it possible to learn from reading narrative about the authenticity of ecology.

In *Ecospatiality* (2021), Lowell Wyse focuses on space to examine prose works in modern and contemporary American literature through the lens of place, analysing texts by William Least Heat-Moon, Willa Cather, Richard Wright and Leslie Marmon Silko, which represent places and the interaction of human and environment. Wyse draws on scholarship in sociology, geography, geocriticism (especially the nonfiction of William Least Heat-Moon) and ecocriticism and shapes a theory of "ecospatiality," a comprehension of place as ecological, spatial and historical.[61] The example or the anecdote opens the book, a strategy like that of some new historicists: "In the opening paragraph of John Cheever's 1964 short story 'The Swimmer,' the protagonist, Neddy Merrill, has an epiphany while lounging by his friends' pool."[62] In discussing literary criticism, Wyse sets out a crucial definition:

> Place, in my formulation, is the convergence of nature, space, and story, and fully appreciating its dynamics is as critical to our interpretation of literary texts as it is to understanding our physical and ethical relationship to the world around us.[63]

Site, sight, story, place, all of which are set in North America, are all important to Wyse's approach in which places in nature come to be places in literature and thus in culture. Wyse also sets his framework by further definitions: "'ecospatial orientation' applies to the ways in which human beings comprehend place—a way of knowing and seeing—as well as to the narrower issue of how place operates in literary texts—a way of reading."[64] Orienting an understanding of place and reading is important. He amplifies: "Just as ecospatiality helps us bridge the conceptual gap between space and nature, geocriticism and ecocriticism, orientation addresses the gap between the physical and the discursive, between nature and culture."[65] Comparison and analogy Wyse employs in analysing the mediation between nature and literature. Reading may help in understanding place and how humans relate to it.[66] For Wyse, scale framing inheres in ecospatial orientation and

> refers to the attentiveness of people and texts to various spatial, temporal, and thematic frames of reference with which they (purport to) make sense of the world. In the spatial sense, scale framing is the attention to various geographical imaginaries, from the local to the global, with infinite spaces in between.[67]

Framing, orientation, reference, the imaginary, space and liminality or threshold all connect here. Wyse discusses Timothy Clark's connection of scale framing to the idea of the Anthropocene and shows how the relation between these two concepts is reforming the conversation of ecocriticism.[68] Moreover, Wyse shifts to topophrenia, which Robert Tally, a theorist of geocriticism, says is a mindfulness of place or "place-mindedness."[69] Wyse analyses how cartographical or spatial scale framing and thematic scale framing (ecological awareness) allow a novel the means of considering place in literature.

Furthermore, Wyse uses an approach or ecospatial orientation that implies an incomplete understanding of a text if it does not account for the intricate dynamics of nature, space and story in regard to text and its cultural context, "the spatiohistorical setting it registers, represents, and refracts."[70] Among other things, Wyse analyses briefly Holling Clancy Holling's *Paddle-to- the-Sea* (1941), a fiction for children that represents the Great Lakes, a book Wyse read in his childhood in Michigan and relates to his own background.[71] He performs an ecospatial reading of twentieth-century American fiction that lends itself to this kind of interpretation, such as Willa Cather's *Death Comes for the Archbishop* (1927), Leslie Marmon Silko's *Ceremony* (1977) and Ana Castillo's *So Far from God* (1993) about New Mexico. Moreover, Wyse is concerned with the watershed of the Salinas Valley in California and examines drought, environmental history and the drilling of wells in John Steinbeck's *East of Eden* (1952) and *To a God Unknown* (1933). More generally, Wyse summarizes the crux:

> In an ecospatial literary criticism, in other words, place matters, both on its own terms and as an underappreciated aspect of literary interpretation. Literature matters, particularly in this context, as a site of ecospatial meaning. And the relationship between them—both the environmentality of the text and the textuality of the environment—matters most of all.[72]

Place matters in literature and literary criticism.

Wyse has some insights into these ecological literary matters. For instance, he observes: "In his nonfictional narratives of place, William Least Heat-Moon explores the identity of American locales, going far beyond mere geographical references to investigate the natural and cultural meanings of places and environments."[73] A wide range of prose works Wyse discusses, examining space to make space to shape his argument on literature, culture and nature. He elaborates: "in *Blue Highways, PrairyErth, and River-Horse,* Heat-Moon uses unique travel experiences derived from cartographic premises to structure his physical, literary, and philosophical investigations of place."[74] Concerning the novels of Central New Mexico by Cather, Silko and Castillo, Wyse stresses their significance as they tell and retell "the stories of communities that are united by geographical, social, and ecological circumstances" while supplementing each other and an understanding of this place and "to read environmentally, to understand the ways that places interact with their own representations" – "new understandings of human beings and the environments they inhabit."[75] This idea of interaction between places of representations of them is intriguing and goes back, in Western texts, to Socrates, Plato and Aristotle about the aesthetics and ethics of mimesis and the representation of reality, not to mention the geographical and ethnological texts among the ancient Greeks and Romans, including Pliny. Wyse moves this debate forward in the context of American literature in the twentieth century in its diversity. Wyse says of Steinbeck: "Reading his Salinas watershed texts through an ecospatial lens reveals both the extent of his literary engagement with his 'native place' and his insistence that we grapple with the intertwined geographical, ecological, and historical dimensions of place itself."[76] Steinbeck is exemplary for literature and literary criticism of place, one that weaves geography, ecology and history. Wyse sees his own reading of Richard Wright's *Native Son* as showing "the extent to which place, rather than being incidental to human experience and fictional narratives, actually structures both in compelling ways."[77] The Afterword has as its subtitle, "An Ethical Orientation," and includes a view of place "as the complex matrix of nature, space, and story that underpins and intertwines with human experience," – "an urgent message in the Anthropocene."[78] Wyse ends with a hope for survival: "To the extent that we comprehend and adopt an ecospatial and multicultural orientation and cultivate an ethic of place, we just might be able to create the kinds of places that allow us to endure."[79] The human and the environment live in a precarious relation, an ecology of vulnerability.

Natural history, ecology and ecospatiality are terms that Chow, Menrisky and Wyse explore, examining American literature and its relation to nature and biology. These authors create studies that invite further thought about the possibilities of ecological literature and criticism, the human relation to the environment. C. P. Snow wrote about the wide gap between two solitudes, two cultures, the scientific and the literary, and declares: "By training I was a scientist: by vocation I was a writer."[80] Snow was unusual. Chow, Menrisky and Wyse largely approach ecology from a literary or humanist point of view. Their language often reflects the professional expression of at least one section of literary criticism or of humanities and social science, whereas scientists might well approach method and style differently. Some of the academic style in this humanistic side of the two cultures since the mid-1960s is not the plain style but is more Ciceronian than Attic. Plain English seems to be part of a humanist past, as, in its most elemental form, for instance, I. A. Richards and C. K. Ogden sought with Basic English. Analytical philosophy and those like Bertrand Russell and Ludwig Wittgenstein sought to understand language but not the language of Shakespeare. Others, such as G. W. F. Hegel and Søren Kierkegaard, Continental philosophers, found power in Shakespeare's words, characters and irony. Perhaps this is a tension between denotative language and poetry (literary language) in our culture. In writing about natural history and ecology, Chow, Menrisky and Wyse navigate the difference between ecological literature and ecological science, which could not be written with such an elaborate style. Whether these three authors' work can reach an audience across the two cultures or to undergraduate students is an open question. Regardless, they are part of literary ecocriticism and its related fields, and their work deserves a place, a time and space, to help us to understand the role of literature in representing and understanding the natural world.

Nature meant and means so much. Growing up, I spent summers in the country, by sea, by lakes, in the mountains. At a primal and visceral level, the destruction of nature feels like the destruction of ourselves and unnatural in all senses. The book began with Darwin and the natural world, and it ends that way. The next chapter contains some concluding thoughts.

Notes

1 See Alexander Menrisky, *Wild Abandon: American Literature and the Identity Politics of Ecology* (Cambridge: Cambridge University Press, 2021); Juliana Chow, *Nineteenth-Century American Literature and the Discourse of Natural History* (Cambridge: Cambridge University Press, 2022); Lowell Wyse, *Ecospatiality: A Place-based Approach to American Literature* (Iowa City: University of Iowa Press, 2021). Thank you to the editors of the journal and to the University of Toronto Press for permission to publish this revised version of Jonathan Locke Hart, "Ecology, Natural History, and Ecospatiality in American Literature," *Canadian Review of American Studies* 53 (3) (2023): 264–275. My thanks to Priscilla L. Walton (in memory) and Christopher Vanderwees for taking an interest in my work generally and for inviting this writing.

2 See Donald H. Pfister, "Asa Gray and Harvard Summer School," *Harvard Papers in Botany* 15 (2) (2010): 305–308.

3 Janet Browne, "Asa Gray and Charles Darwin: Corresponding Naturalists," *Harvard Papers in Botany* 15 (2) (December 2010): 210.

4 Browne, "Asa Gray and Charles Darwin," 211. Asa Gray, letter from Charles Darwin, 21 December 1859, S2592, Library, Gray Herbarium, Harvard University. See Charles Darwin, *On the Origin of Species by Means of Natural Selection, or the Preservation of Favoured Races in the Struggle for Life* (London: John Murray, 1859).

5 See Anderson Hunter Dupree, *Asa Gray: American Botanist, Friend of Darwin*, reprint of 1959 ed. (Baltimore: Johns Hopkins University Press, 1988).

6 Jaboury Ghazoul, *Ecology: A Very Short Introduction* (Oxford: Oxford University Press, 2020).
7 Charles Elton qtd in Ghazoul, *Ecology*, 10.
8 Ghazoul, *Ecology*, 10.
9 Ghazoul, *Ecology*, 10.
10 Ghazoul, *Ecology*, 118–129, see Fikret Berkes, *Sacred Ecology*, 3rd edition (New York and London: Routledge, 2012).
11 Ghazoul, *Ecology*, 130.
12 Ghazoul, *Ecology*, 147–148.
13 (*OED* Online). "natural history, n.". OED Online. March 2023. Oxford University Press.
14 (*OED* Online). "natural history, n.". OED Online. March 2023. Oxford University Press.
15 Chow, *Nineteenth-Century*. See Janet Browne, "Biogeography and Empire," in *Cultures of Natural History: From Curiosity to Crisis*, edited by Nicholas Jardine, James Secord, and Emma Spary (Cambridge: Cambridge University Press, 1996), 305–321.
16 Chow, *Nineteenth*-Century, 1.
17 Chow, *Nineteenth-Century*, 2.
18 Chow, *Nineteenth-Century*, 2, see Laura Dassow Walls, *The Passage to Cosmos: Alexander von Humboldt and the Shaping of America* (Chicago, IL: University of Chicago Press, 2009); Aaron Sachs, *The Humboldt Current: Nineteenth-Century Exploration and the Roots of American Environmentalism* (New York: Penguin Books, 2006). See also Lawrence Buell, *The Environmental Imagination: Thoreau, Nature Writing, and the Formation of American Culture* (Cambridge, MA: Harvard University Press, 1995); Lawrence Buell, *The Future of Environmental Criticism: Environmental Crisis and Literary Imagination* (Malden, MA: Blackwell, 2005).
19 Chow, *Nineteenth*-Century, 4.
20 Chow, *Nineteenth-Century*, 11.
21 Chow, *Nineteenth-Century*, 11.
22 Chow, *Nineteenth-Century*, 14.
23 Chow, *Nineteenth*-Century, 15.
24 Chow, *Nineteenth-Century*, 31–68.
25 Chow, *Nineteenth-Century*, 78.
26 Chow, *Nineteenth-Century*, 96.
27 Chow, *Nineteenth-Century*, 98.
28 Chow, *Nineteenth-Century*, 141.
29 Chow, *Nineteenth-Century*, 142.
30 Chow, *Nineteenth-Century*, 180.
31 Chow, *Nineteenth-Century*, 181.
32 Chow, *Nineteenth*-Century, 182.
33 See Chow, *Nineteenth-Century*, vi.
34 Chow, *Nineteenth-Century*, 183.
35 Chow, Nineteenth-Century, 189.
36 Chow, *Nineteenth-Century*, 190.
37 "trope, n." OED Online. March 2023. Oxford University Press.
38 "trope, n." *OED* Online.
39 "ecology," *OED*). "ecology, n.". OED Online. March 2023. Oxford University Press.
40 Menrisky, *Wild Abandon*, 1.

41 Menrisky, *Wild Abandon,* 1.
42 Menrisky, *Wild Abandon,* 2.
43 Menrisky, *Wild Abandon,* 3.
44 Menrisky, *Wild Abandon,* 5.
45 Menrisky, *Wild Abandon,* 13.
46 Menrisky, *Wild Abandon,* 16.
47 Menrisky, *Wild Abandon,* 17.
48 Menrisky, *Wild Abandon,* 27–58
49 Menrisky, *Wild Abandon,* 29.
50 Menrisky, *Wild Abandon,* 59–94.
51 Menrisky, *Wild Abandon,* 86.
52 Menrisky, *Wild Abandon,* 95–132.
53 Menrisky, *Wild Abandon,* 98.
54 Menrisky, *Wild Abandon,* 133–169.
55 Menrisky, *Wild Abandon,* 135.
56 Menrisky, *Wild Abandon,* 167.
57 Menrisky, *Wild Abandon,* 170–198.
58 Menrisky, *Wild Abandon,* 170.
59 Menrisky, *Wild Abandon,* 199–209.
60 Menrisky, *Wild Abandon,* 209.
61 Wyse, *Ecospatiality,* 3.
62 Wyse, *Ecospatiality,* 1.
63 Wyse, *Ecospatiality,* 3–4.
64 Wyse, *Ecospatiality,* 4.
65 Wyse, *Ecospatiality,* 4.
66 Wyse, *Ecospatiality,* 4.
67 Wyse, *Ecospatiality,* 4–5.
68 Wyse, *Ecospatiality,* 5; see Timothy Clark, *Ecocriticism on the Edge: The Anthropocene as a Threshold Concept* (London: Bloomsbury Academic, 2015).
69 Wyse, *Ecospatiality,* 5; see Robert T. Tally, *Topophrenia: Place, Narrative, and the Spatial Imagination* (Bloomington: Indiana University Press, 2019), 30; see also Robert T. Tally Jr., *Spatiality* (New York: Routledge, 2013).
70 Wyse, *Ecospatiality,* 6, see 5.
71 Wyse, *Ecospatiality,* 19–26.
72 Wyse, *Ecospatiality,* 31.
73 Wyse, *Ecospatiality,* 33.
74 Wyse, *Ecospatiality,* 89.
75 Wyse, *Ecospatiality,* 127.
76 Wyse, *Ecospatiality,* 168.
77 Wyse, *Ecospatiality,* 172.
78 Wyse, *Ecospatiality,* 210.
79 Wyse, *Ecospatiality,* 210.
80 Charles Percy Snow, *The Two Cultures* (Cambridge: Cambridge University Press, 1998), 1.

Conclusion

This study is comparative and involves works about nature, science, literature and culture in the North Atlantic world of Britain, the United States, France and Canada. Natural history, botany, poetry, physics, mathematics, logic, computers, history, literature, imperialism, influence, space, Indigenous culture, ecology and other subjects are all part of the core of this book. I long admired, among other scholarship, the work of Northrop Frye and Harry Levin in comparative literature and Anthony Pagden in comparative history.[1] My own journey has been one in which I am of no school, and I have often found the diverse work of others to be helpful, insightful, suggestive and even inspiring. In *Grounds of Comparison*, Levin makes an observation on his own method:

> I have gone out of my way to avoid identification with any particular school or coterie or set of dogmas. It has always seemed to me that, if it is truth we seek, then every conceivable path to it should be considered viable, and that the seeker should try as many of them as he can, or at least do his best to keep them open.[2]

To seek truth and the many paths and openness and flexibility in that search gives a way forward.

Albert Einstein's speech, "Science and Civilization," was given at Royal Albert Hall on 3 October 1933, a clip of which is still available to watch.[3] Einstein says:

> Let us hope that a historian delivering judgment in some future period when Europe is politically and economically united, will be able to say that in our days the liberty and honour of this Continent was saved by its Western nations, which stood fast in hard times against the temptations of hatred and oppression; and that Western Europe defended successfully the liberty of the individual, which has brought us the

DOI: 10.4324/9781003651871-16

advance of knowledge and invention – liberty without which life to a self-respecting man is not worth living.[4]

Einstein imagines a historian in a future united Europe looking back on the crisis in 1933 and asserts that individual liberty has advanced invention and knowledge. My book focuses mainly on remarkable individuals in families and society. Their freedom of thought and self-expression are distinctive and helped to move society forward. Einstein knows what the crux is: "If we want to resist the powers which threaten to suppress intellectual and individual freedom we must keep clearly before us what is at stake" and soon adds: "Without such freedom there would have been no Shakespeare, no Goethe, no Newton, no Faraday, no Pasteur and no Lister."[5] Einstein stresses that to resist the suppression of freedom, we must know the stakes, something he elaborates at length. Freedom allows for the great figures in science and literature, and Einstein uses an anaphoric "no," thereby showing the negation or lack of culture without the freedom of individuals. The struggle is never done, and, as Einstein realized in 1933, it was important to value freedom to resist powers threatening to suppress it. This speech is still germane. One of the principal aims of my book is to show the expression and self-expression of scientists, artists, humanists and others who stand up for intellectual and individual freedoms. In a community of the free, their knowledge and invention inform their art and science, their words and thoughts.

Science and reason are robust but in a precarious situation. Ignorance, irrationality, dogma, prejudice and folly press them on all sides. Often until well into the twentieth century, science education took a back seat to humanities, which in recent decades have been pushed into the background. Charles Darwin and Pierre Curie, as discussed in my book, were thought to be slow, and how wrong it was and is to expect one kind of mind. Both had to learn languages and classics, and both observed nature and thought deeply. Darwin's own words and Marie Curie's representation of Pierre reveal the development of their minds. Like Bertrand Russell and Alan Turing, they had unconventional minds, as did Einstein. Schools can have a hard time coping with such independence and unconventional ways of seeing and knowing. The Grays and the Thaxters had their own ways and styles. The same concerns about being slow dogged the young Winston Churchill. The imperial and anti-imperial and the rights and dignity of humans are concerns that make their way into those I have discussed, for those who focus on nature and culture. A shared humanity and equality of opportunity are ideas worth pursuing still.

Ancient Greek was compulsory for admission to University of Cambridge until 1919 but had been debated since 1870.[6] Imagine if it had been advanced mathematics in its place. The scientists trained before the First World War were cultured and often trained in classics, as we can see from

Darwin to Russell. A Prime Minister, William Gladstone, was trained in mathematics and classics at Oxford and was multilingual. H. H. Asquith, also trained at Oxford, was conversant in the classics and became a lawyer. As Prime Minister, Asquith addressed the Classical Association of England on 9 October 1908 in Birmingham. Here is part of what he said:

> There may be as much pedantry and waste of time in wrangling over the question to which of our nine hypothetical Minoan epochs a particular potsherd belongs, as in elaborating theories about the different usages of ἄν and οὖν. The shadow of the commentator, whatever may be his particular calling – textual criticism, grammar, excavation – should never be allowed (as it so often has been) to obscure and almost to obliterate the writing of genius. The true scholar values and uses all these aids and lights, each in its due proportion; but the true scholar is rare.[7]

This is not too different from what Levin says. Like Russell and Asquith, Levin favours clarity and open-mindedness and not obscurantism, pedantry and dogma. Teachers and researchers need to be open to difference and not be close-minded. Slow and deep observation of nature and slow and deep reading are gifts a student may have, and gifts they are. Marie Curie praised Pierre Curie's parents for their understanding of their son and what he needed for education. Darwin and Curie loved to be out in nature observing.

And so, it is incumbent on humans to be in and understand nature. Humanists, poets, artists are not exempt. My book is an attempt to understand nature and human nature, how humans look at nature and themselves. These figures who interpret nature, humanity and themselves are part of a quest for truth and a sense of reality, however they are like a firefly over a marsh, receding. Perhaps that quest is quixotic, foolish or refractory, but however asymptotic the journey is, it is worth the trouble and the beauty. Perhaps with Hamlet, we might say:

> There is a special providence in the fall of a sparrow. If it be [now], 'tis not to come; if it be not to come, it will be now; if it be not now yet it [will] come – the readiness is all. Since no man, of aught he leaves, knows what is't to leave betimes. Let be.
>
> (*Hamlet*, V.ii. 219–224)[8]

Hamlet is meditating on Providence, death and its relation to knowledge and time. As humans do not know what is left behind at death, then what is it to them to leave early? This view of time in connection with the human is one way to regard the temporality of the world. And I conclude with Einstein. Michele Besso and he were friends since they were patent clerks in Berne. Besso died in March 1955, and Einstein wrote a letter in German to his family days later, and weeks before Einstein's own death, that ends thus:

> Now he has again preceded me a little in parting from this strange world. This has no importance. For people like us who believe in physics, the separation between past, present and future has only the importance of an admittedly tenacious illusion.[9]

Thoughts on time, life and death from Shakespeare and Einstein open vistas, each from a distinct human, one writing a fiction and thus an illusion, another thinking of time as an illusion for a physicist. As Einstein said before, the freedom of the individual allows for Shakespeare, but it also allows for Einstein.

Notes

1 Northrop Frye, *Anatomy of Criticism : Four Essays*. First Princeton classics edition (new foreword by David Damrosch) (1957; Princeton: Princeton University Press, 2020); Harry Levin, *The Implications of Literary Criticism*, ed. Jonathan Hart (Paris: H. Champion, 2011), (Bibliothèque de littérature générale et comparée 96.); Anthony Pagden, *The Fall of Natural Man: The American Indian and the Origins of Comparative Ethnology* (Cambridge: Cambridge University Press, 1982). David Damrosch's new foreword is a good analysis of *Anatomy* and its influence and afterlife – Damrosch concludes: "At the heart of its interlocking cycles of myths, modes, symbols, and archetypes, *Anatomy of Criticism* advances a poetics of liberation" (xvii). Here, we have freedom again. See Jonathan Hart, *Northrop Frye: The Theoretical Imagination* (London: Routledge, 1994); Jay Garcia, Review of *The Explanatory Critic: Harry Levin and Literary Criticism*, by Harry Levin (edited by Jonathan Hart). *Comparative Literature Studies* 51 (3) (2014): 491–503. I thank and remember my teacher, Northrop Frye, and my mentor Harry Levin and thank Anthony Pagden, my supervisor at Cambridge. I owe a great deal to them and have learned a great deal from them, and I also wish to express my gratitude for my years at Toronto, Harvard and Cambridge. In thanks, I have written about or drawn on Frye, Levin and Pagden in various of my works over the years.

2 Harry Levin, *Grounds for Comparison* (Cambridge: Harvard University Press, 1972), 4.

3 Albert Einstein's speech at Royal Albert Hall in protest against Nazi policy (1933), British Pathé; 1 minute and fifteen seconds from the beginning of the speech, delivered in English, see, for instance, https://www.britishpathe.com/asset/139299/.

4 Albert Einstein, "Programme, Royal Albert Hall, October Third ... 1933 ... Organized by the Refugee Assistance Fund ... Professor Einstein's Speech ... Science and Civilization," *Special Collections & Western MSS, Bodleian Library*, Oxford University Library Services, [n.p.].

5 Einstein, "Programme ... Science and Civilization ...," [n.p.].

6 See, for instance, J. K. Stephen, *The Living Languages: A Defence of the Compulsory Study of Greek at Cambridge* (Cambridge: Macmillan and Bowes, 1891); Christopher Stray, *Classics Transformed: Schools, Universities, and Society in England, 1830–1960* (Oxford: Clarendon Press, New York: Oxford University Press, 1998); Judith Raphaely, "Nothing but gibberish and shibboleths? The Compulsory Greek debates, 1870–1919," in *Classics in 19th and 20th Century Cambridge: Curriculum, Culture and Community*, edited by C.A. Stray (Cambridge Philological Society, Suppl. Vol. 24, 1999), 71–94; J. K. Stephen, "The Living Languages: Why Cambridge Needs Greek," ANTIGONE. The journal says about itself: "Antigone is a New and Open Forum for Classics in the Twenty-first Century." See https://antigonejournal.com/2025/05/the-living-languages-compulsory-greek-j-k-stephen/.

7 ADDRESS TO THE CLASSICAL ASSOCIATION OF ENGLAND from H.H. Asquith, Prime Minister of the United Kingdom Birmingham, 9 October 1908, reprinted in *Antigone*, https://antigonejournal.com/2025/07/asquith-on-the-classics/.

8 William Shakespeare, *The Riverside Shakespeare: The Complete Works*. 2nd ed., edited by G. Blakemore Evans with J. J. M Tobin (Boston: Houghton Mifflin, 1997), 1231.

9 The translation is from "CHRISTIE'S Time's arrow: Albert Einstein's letters to Michele Besso Thomas Venning, Head of Books & Manuscripts at Christie's in London, explores the touching correspondence between Einstein and his dear friend of more than 50 years – illustrated with letters offered in past and upcoming sales," © CHRISTIE'S 2025; see https://www.christies.com/en/stories/einstein-letters-to-michele-besso-3ecca787d9be4463a8695269259d6c00. See also the letter published (in French and German) in Pierre Speziali (ed. and tr.). *Albert Einstein. Michele Besso. Correspondance 1903–1955* (Paris: Hermann, 1972) No. 211.

Index

Note: Page numbers followed by "n" refer to end notes.